Vibration Reduction for Flexible Structures by Smoothing Commands

Jie Huang

Lulu Press

627 Davis Drive, Suite 300,

Morrisville, NC 27560

www.lulu.com

Edition: January, 2020

ISBN 978-0-244-49523-7

Jie Huang

School of Mechanical Engineering

Beijing Institute of Technology

5 South Zhongguancun Street, Haidian District, Beijing, China, 100081

Preface

This book described vibration reduction for flexible structures by the command smoothing techniques. The command smoothing technique is a kind of open-loop controller. The operator's commands filter through a piecewise continuous function, called smoother, to produce smoothed commands. Smoothed commands move flexible dynamic systems toward desired positions with minimal oscillations.

Five types of command smoother were reported including one-piece smoother, two-pieces smoother, three-pieces smoother, four-pieces smoother, and the smoother for multi-mode Duffing oscillators. All smoothers are robust to changes in the system parameters and working conditions.

The command smoothing technique has been successfully applied to industrial bridge and tower cranes, sloshing suppression in a moving liquid container, flexible link manipulators, high-speed cam and follower systems, and helicopters slung loads.

Contents

Chapter 1. Introduction to Control of Flexible Structures

1.1 Vibrations of Flexible Structures

Numerous dynamic systems with flexible structures suffer from undesirable transient and residual vibrations. Detrimental effects will cause serious problems for positioning accuracy, operating speed, effectiveness, and safety [1]. Therefore, there is a need to study dynamics and control of flexible structures so that unwanted vibrations can be effectively reduced for safe and effective operations.

1.2 Control of Flexible Structures

Many scientists have worked to provide solutions to the challenging problem posed by the flexible structure. The work can roughly be broken into two categories: feedback control, and open-loop control. The feedback control strategies use measurement and estimation of vibrational state to suppress oscillations in a closed loop [2-3], such as proportional-integral-derivative control, sliding-mode control, adaptive control, optimal control, H_∞ control, and fuzzy control. However, accurately sensing vibrations is difficulty toward the application of the feedback controller. Meanwhile, the conflict between the computer-based feedback controller and actions of human operator is also an obstacle.

Open-loop control schemes modify the input to create prescribed motions that cause minimal vibrations. Input shaping is a kind of open-loop controller. It can effectively suppress oscillations for many types of flexible dynamic systems including bridge cranes [4], tower cranes [5], boom cranes [6], container cranes [7], coordinate measurement machines [8], space crafts [9], robotic arms [10], robotic work cells [11], demining robots [12], micro-milling machines [13], nano-positioning stages [14], and linear step motors [15]. The input shaping process is demonstrated as follows. The original command produces an oscillatory response. To eliminate the oscillatory response, the original command is convolved with a series of impulses, called the input shaper, to create a shaped command. The shaped command can move the flexible dynamic system without inducing vibrations. The convolution is performed by simply multiplying the original command by the amplitude of the first impulse, and adding it to the original command multiplied by the amplitude of the other impulses and shifted in time by the corresponding delay period.

Hundreds of papers reported smooth command profiles to move flexible dynamic systems for vibration reduction, such as S-curves [16], trigonometric transition functions [17], Gaussians [18], spline function [19] and cam polynomials [20]. The tendency for exciting vibrations of the flexible system has been reduced by using smooth profiles. Ramping up to peak acceleration and ramping down to constant velocity produce a smooth velocity trajectory. The smooth transitions between boundary conditions avoid vibrations. Vibration reduction comes at the cost of increased duration of this ramp-up time. While significant work has been directed at smooth profiles to reduce vibrations, however, these methods usually fail to fully exploit the known property of flexible structures. Instead, they simply provide a low-pass filtering effect [21].

1.3 Brief of Command Smoothing

Command smoothing techniques filter operator's commands to move flexible dynamic systems with minimal oscillations. The operator's command is sent through a smoother for suppressing oscillations, thereby producing a smoothed command, which moves flexible structures toward desired positions with minimum oscillations. The smoother is a piecewise continuous profile as a function of natural frequency and damping ratio of flexible structures. The command smoothing technique has been successfully applied to industrial cranes, liquid container transportation, a flexible manipulator, and a high-speed cam.

1.4 Chapter Summary

This chapter overviews the vibration and control of the flexible structure, and briefly introduces the command smoothing technique and its applications.

Chapter 2. Command Smoothing Techniques

Smoothing driving commands can suppress vibrations of the flexible structure caused by commanded motions. The driving command filters through a smoother to create a smoothed command, which moves flexible structures towards the desired position with minimum vibrations. The smoother inherent in the limited response of flexible structures causes a limited response to specified frequencies. The smoother is designed by estimating natural frequencies and damping ratios of flexible structures.

2.1 One-Piece Smoother

2.1.1 Design of One-Piece Smoother

The command smoother is a piecewise function, s_1. If the flexible structure can be modeled as a second-order harmonic oscillator, then the response resulting from the piecewise function, s_1, is:

$$f(t) = \int_{\tau=0}^{+\infty} s_1(\tau) \frac{\omega}{\sqrt{1-\zeta^2}} e^{-\zeta\omega(t-\tau)} \sin(\omega(t-\tau)\sqrt{1-\zeta^2})d\tau, \tag{2.1}$$

where ω and ζ are the natural frequency and damping ratio of the system. The vibrational amplitude of response (2.1) is:

$$A(t) = \frac{\omega}{\sqrt{1-\zeta^2}} e^{-\zeta\omega t} \sqrt{[S(\omega,\zeta)]^2 + [C(\omega,\zeta)]^2}, \tag{2.2}$$

where,

$$S(\omega,\zeta) = \int_{\tau=0}^{+\infty} s_1(\tau) e^{\zeta\omega\tau} \sin(\omega\tau\sqrt{1-\zeta^2})d\tau, \tag{2.3}$$

$$C(\omega,\zeta) = \int_{\tau=0}^{+\infty} s_1(\tau) e^{\zeta\omega\tau} \cos(\omega\tau\sqrt{1-\zeta^2})d\tau. \tag{2.4}$$

If equations (2.3) and (2.4) are limited to zero, the piecewise function, s_1, would cause zero residual vibrations. Then resulting from equations (2.3) and (2.4), the smoother is described by [22]:

$$s_1(\tau) = \begin{cases} u_1 e^{-\zeta\omega\tau}, & 0 \leq \tau \leq T_1 \\ 0 & \text{others} \end{cases}, \tag{2.5}$$

where T_1 is a damped oscillation period, and the coefficient, u_1 is the value of the piecewise function, s_1, at time zero. The equation (2.5) is a continuous

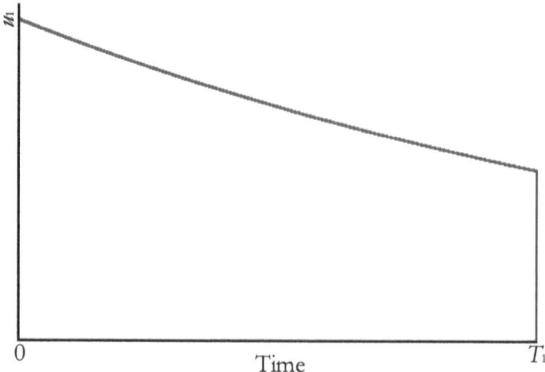

Figure 2.1 Curve of one-piece smoother.

piecewise function with one piece, and so is called one-piece smoother. The curve of one-piece smoother is shown in Figure 2.1, which is a function of natural frequency and damping ratio.

Another constraint should be applied to ensure that the smoothed command reaches the same set-point as the original command. The shaping process is to have unity gain in order to satisfy this requirement. Then, the integral of the piecewise function, s_1, is limited to one:

$$\int_{\tau=0}^{+\infty} s_1(\tau)d\tau = 1. \tag{2.6}$$

Resulting from the equation (2.5) and constraint (2.6), the coefficient, u_1, can be obtained:

$$u_1 = \frac{\zeta\omega}{(1-M_1)}, \tag{2.7}$$

where

$$M_1 = e^{-2\pi\zeta/\sqrt{1-\zeta^2}}. \tag{2.8}$$

Resulting from equations (2.5) and (2.7), the transfer function of the one-piece smoother is given by [22]:

$$s_1(s) = \frac{\zeta\omega(1-M_1 e^{-2\pi s/(\omega\sqrt{1-\zeta^2})})}{(1-M_1)(s+\zeta\omega)}. \tag{2.9}$$

By using the one-piece smoother (2.9), the vibration amplitude (2.2) will be limited to zero for any arbitrary command. This smoothing process is

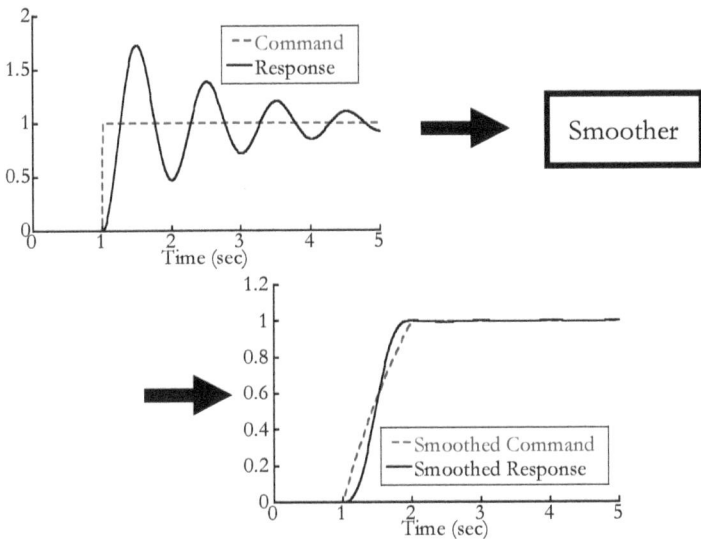

Figure 2.2 Command smoothing process.

demonstrated in Figure 2.2. The original command produces an oscillatory response represented by the solid line labeled "Response". To eliminate the oscillatory response, the original command filters through the smoother to create the smoothed command shown at the bottom of Figure 2.2. The smoothed command can move the flexible dynamic systems without inducing vibrations.

2.1.2 Sensitivity to Modeling Error

The frequency and damping ratio of flexible structures may not be known accurately in many cases. It then becomes important to evaluate how this uncertainty can translate into the percentage residual amplitude (PRA). The vibration amplitude at time zero is:

$$A_0 = \frac{\omega}{\sqrt{1-\varsigma^2}} \ . \tag{2.10}$$

Dividing equation (2.2) by equation (2.10) yields the percentage residual amplitude:

$$PRA = e^{-\varsigma \omega t} \sqrt{S(\omega,\varsigma)^2 + C(\omega,\varsigma)]^2} \ . \tag{2.11}$$

Figure 2.3 shows the frequency sensitivity curve for the one-piece smoother designed for a zero damping ratio. The width of each curve, which lies below a specified vibrational level, is defined as the insensitivity.

5

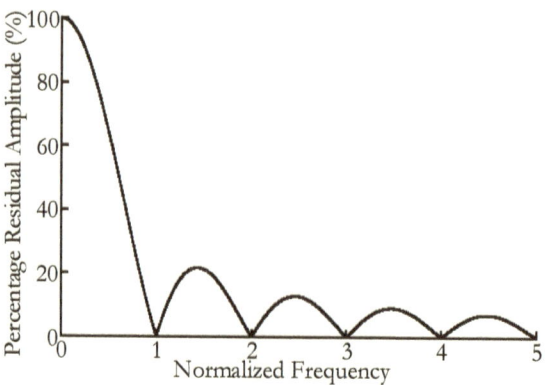

Figure 2.3 Frequency sensitivity for one-piece smoother.

It provides a quantitative measure of the robustness. Less than 5% of percentage residual amplitude is the acceptable level of vibrations. The 5% insensitivity of the one-piece smoother ranges from 0.953 to 1.052. The normalized frequency (ω/ω_m) in this section is defined as the ratio of the real frequency to the modeled frequency, where ω_m is the modeled frequency. The percentage residual amplitude is zero at the modeling frequency ($\omega/\omega_m=1$), while the sensitivity curve reaches a maximum of 100% at the zero normalized frequency. In addition, the percentage residual amplitude for the one-piece smoother at the integer normalized frequency is also zero. Increasing normalized frequency will decrease the peak of percentage residual amplitudes for the one-piece smoother. Additionally, the modeling error in the damping has few impacts on the residual vibration amplitudes.

2.2 Two-Pieces Smoother

2.2.1 Design of Two-Pieces Smoother

A new constraint must be added to increase the robustness of the smoother under variations of the frequency and damping ratio. The derivative of equations (2.3) and (2.4) with respect to ω and ζ should also be set equal to zero. Then changes in the frequency and damping ratio will result in small changes in the residual vibration amplitude. The following equations should be applied in order to satisfy this requirement:

$$\int_{\tau=0}^{+\infty} \tau \cdot s_2(\tau) e^{\zeta\omega\tau} \sin(\omega\tau\sqrt{1-\zeta^2}) \mathrm{d}\tau = 0, \qquad (2.12)$$

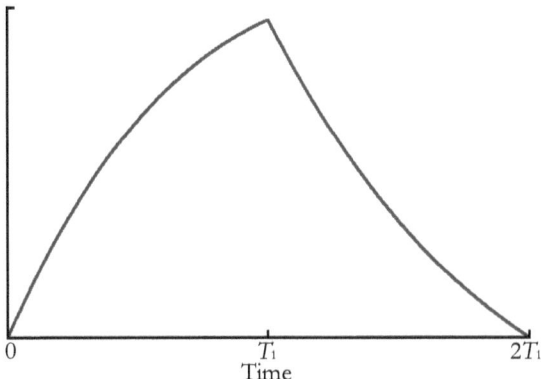

Figure 2.4 Curve of two-pieces smoother.

$$\int_{\tau=0}^{+\infty} \tau \cdot s_2(\tau) e^{\zeta \omega \tau} \cos(\omega \tau \sqrt{1-\zeta^2}) \mathrm{d}\tau = 0, \qquad (2.13)$$

where s_2 is the two-pieces smoother.

Limiting equations (2.3) and (2.4) to zero and resulting from constraints (2.12) and (2.13), the two-pieces smoother is given by [23]:

$$s_2(\tau) = \begin{cases} \tau u_2 e^{-\zeta \omega \tau}, & 0 \le \tau \le T_1 \\ (2T_1 - \tau)u_2 e^{-\zeta \omega \tau}, & T_1 < \tau \le 2T_1, \\ 0, & \text{others} \end{cases} \qquad (2.14)$$

where u_2 is the coefficient of the two-pieces smoother. The equation (2.14) is a continuous piecewise function with two pieces, and so is called two-pieces smoother. Resulting from the unity-gain constraint (2.6) for the smoother (2.14), the coefficient, u_2, has the form:

$$u_2 = \frac{\zeta^2 \omega^2}{(1-M_1)^2}. \qquad (2.15)$$

The curve of two-pieces smoother, which is a piecewise continuous profile as a function of natural frequency and damping ratio, is shown in Figure 2.4. Resulting from equation (2.14), the transfer function of the two-pieces smoother is described by [23]:

$$s_2(s) = \frac{\zeta^2 \omega^2}{(1-M_1)^2} \cdot \frac{(1-M_1 e^{-T_1 s})^2}{(s+\zeta \omega)^2}. \qquad (2.16)$$

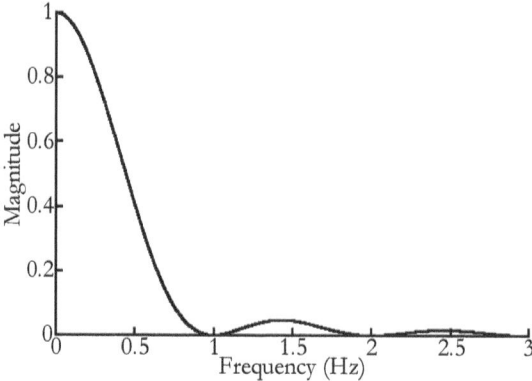

Figure 2.5 Magnitude of Laplace transform of two-pieces smoother.

When the modeled frequency and damping ratio are correct, the vibration amplitude (2.2) will be limited to zero for any arbitrary command by using the two-pieces smoother (2.14). The magnitudes of the two-pieces smoother for a damped natural frequency of 1Hz are shown in Figure 2.5. The smoother is a combination of multi-notch and low-pass filters. It is apparent that the smoother cannot excite flexible dynamic systems which have modes corresponding to the notch frequencies at 1, 2, 3 Hz, etc. It is also clear that the smoother cannot excite high frequencies. The magnitude of the smoother is limited to one at the zero frequency. This is because the smoothing process is to have unity gain.

The smoothing process increases the rise time by the duration of the two-pieces smoother. The duration of the two-pieces smoother, R_{s2}, is twice as long as a modeled damped vibration period:

$$R_{s2} = 2T_1.$$

(2.17)

2.2.2 Higher-Order Derivative Smoother

The higher-order derivatives of equations (2.3) and (2.4) with respect to ω and ζ will add more robustness of the smoother. This process has the general form. Limiting the nth derivatives of equations (2.3) and (2.4) with respect to ω and ζ to zero yields:

$$\int_{\tau=0}^{+\infty} \tau^n s_{nD}(\tau)e^{\zeta\omega\tau} \sin(\omega\tau\sqrt{1-\zeta^2})\mathrm{d}\tau = 0 ,$$

(2.18)

$$\int_{\tau=0}^{+\infty} \tau^n s_{nD}(\tau)e^{\zeta\omega\tau} \cos(\omega\tau\sqrt{1-\zeta^2})\mathrm{d}\tau = 0 ,$$

(2.19)

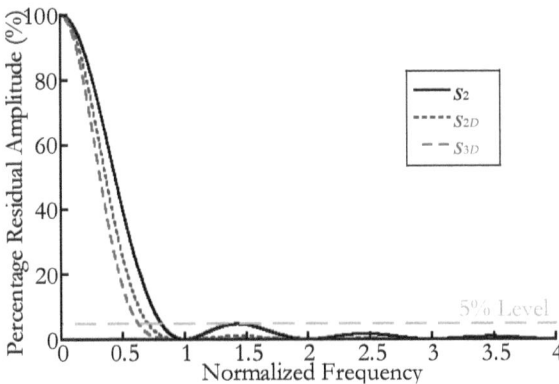

Figure 2.6 Frequency sensitivity for high-order derivative smoother.

where s_{nD} is the smoother with higher-order derivative. Limiting equations (2.3) and (2.4) to zero and resulting from the unity-gain constrain and constraints (2.18) and (2.19), the transfer function of the smoother with the nth derivative (s_{nD}) with respect to ω and ζ is given by:

$$s_{nD}(s) = \frac{\zeta^{(n+1)}\omega^{(n+1)}}{(1-M_1)^{(n+1)}} \cdot \frac{(1-M_1 e^{-T_1 s})^{(n+1)}}{(s+\zeta\omega)^{(n+1)}} . \tag{2.20}$$

The duration of equation (2.20), R_{snD}, is (n+1) times as long as a modeled damped vibration period:

$$R_{snD} = (n+1)T_1 . \tag{2.21}$$

2.2.3 Sensitivity to Modeling Error

Figure 2.6 shows the frequency sensitivity curve for the two-pieces smoother designed for a zero damping ratio. The percentage residual amplitude is limited to zero when the modeled frequency is correct. The zero-slope at the design frequency is resulting from the derivative constraints (2.18) and (2.19). The 5% insensitivity of the smoother is from 0.81 to infinity. The smoother has more insensitivity at higher frequencies. As the normalized frequency increases, peaks of the percentage residual amplitude for the smoother will decrease. Therefore, the two-pieces smoother can reduce high-frequency vibrations. This performance benefits vibration suppression for multi-mode systems. While the two-pieces smoother is designed to suppress first-mode vibrations, it will also reduce high-mode vibrations. Meanwhile, the design of the smoother does not need to estimate high-mode frequencies.

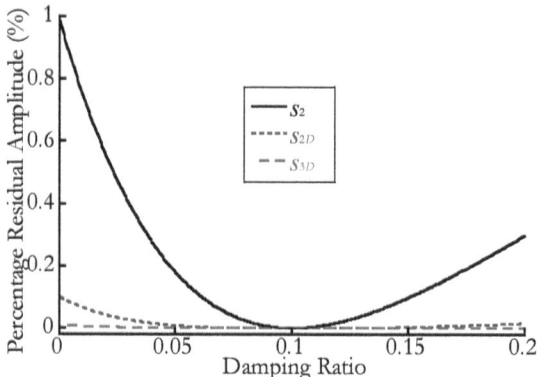

Figure 2.7 Damping sensitivity for high-order derivative smoother.

Figure 2.6 also shows frequency sensitivity curves for the smoother with the second derivative, and the smoother with the third derivative, designed for a zero damping ratio. The 5% frequency insensitivity of the smoother with the second derivative varies from 0.7 to infinity, and that with the third derivative is from 0.63 to infinity. Therefore, taking additional derivatives increases the frequency insensitivity.

Figure 2.7 shows the damping sensitivity curve for the two-pieces smoother designed for a modeled damping ratio of 0.1. The robustness to changes in the damping ratio follows similar trends to that in the frequency. There is one difference that the damping ratio is not normalized. That is because small changes in the damping ratio result in large changes in the normalized damping ratio. When the modeled damping ratio is correct, the percentage residual amplitude is also zero. The derivative constraints also cause a zero-slope performance at the modeling damping ratio. A dramatic reduction in the residual vibration amplitude exists for all values of damping shown in Figure 2.7. Moreover, the insensitivity of the two-pieces smoother tolerates extremely large variations in the damping. Variations in the damping ratio do not have a large effect on residual vibrations as changes in the frequency.

Figure 2.7 also shows damping sensitivity curves for the smoother with the second derivative, and the smoother with the third derivative, designed for a modeled damping ratio of 0.1. Taking additional derivatives will also increase the damping insensitivity. The price for each additional derivative is an increase in the smoothing duration by a modeled damped vibration period.

2.3 Three-Pieces Smoother

2.3.1 Design of Three-Pieces Smoother

If equations (2.3) and (2.4) are limited to zero, the smoother would cause zero residual vibrations. Nevertheless, some degree of uncertainty, which can result from poorly known or time-varying parameters and nonlinearities of the system, occurs in an actual system. Consequently, the smoother could not limit vibrations to zero on a practical system. Therefore, vibrations at the design point could be suppressed to a tolerable level:

$$[e^{-\zeta \omega R_{s3}} \sqrt{[S(\omega,\zeta)]^2 + [C(\omega,\zeta)]^2}\,]_{\substack{\omega=\omega_m \\ \zeta=\zeta_m}} = V_{tol}\,, \tag{2.22}$$

where ω_m is the modeled frequency, ζ_m is the modeled damping ratio, R_{s3} is the duration of the three-pieces smoother, s_3, and V_{tol} is the tolerable level of vibrations.

Additionally, the smoother should have well robust to the system parameter. Forcing vibrations at the modified frequencies, $p \cdot \omega_m$ and $r \cdot \omega_m$, to zero, could maximize the robustness to modeling errors in the frequency. The p and r are the amending coefficients. The zero-vibration constraints at the modified frequencies, p and r, are given by:

$$[\int_{\tau=0}^{+\infty} s_3(\tau)e^{\zeta\omega\tau}\sin(\omega\sqrt{1-\zeta^2}\,\tau)\mathrm{d}\tau]_{\substack{\omega=p\omega_m \\ \zeta=\zeta_m}} = 0\,, \tag{2.23}$$

$$[\int_{\tau=0}^{+\infty} s_3(\tau)e^{\zeta\omega\tau}\cos(\omega\sqrt{1-\zeta^2}\,\tau)\mathrm{d}\tau]_{\substack{\omega=p\omega_m \\ \zeta=\zeta_m}} = 0\,, \tag{2.24}$$

$$[\int_{\tau=0}^{+\infty} s_3(\tau)e^{\zeta\omega\tau}\sin(\omega\sqrt{1-\zeta^2}\,\tau)\mathrm{d}\tau]_{\substack{\omega=r\omega_m \\ \zeta=\zeta_m}} = 0\,, \tag{2.25}$$

$$[\int_{\tau=0}^{+\infty} s_3(\tau)e^{\zeta\omega\tau}\cos(\omega\sqrt{1-\zeta^2}\,\tau)\mathrm{d}\tau]_{\substack{\omega=r\omega_m \\ \zeta=\zeta_m}} = 0\,, \tag{2.26}$$

where s_3 is the three-pieces smoother.

A zero-slope constraint should be added for increasing the robustness. Then the derivative of equation (2.22) with respect to frequency should be limited to zero:

$$[\frac{\mathrm{d}[(e^{-\zeta\omega R_{s3}}S(\omega,\zeta))^2 + (e^{-\zeta\omega R_{s3}}C(\omega,\zeta))^2]}{\mathrm{d}\omega}]_{\substack{\omega=\omega_m \\ \zeta=\zeta_m}} = 0\,. \tag{2.27}$$

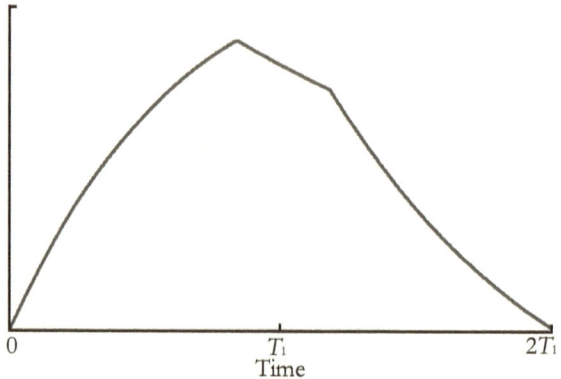

Figure 2.8 Graph illustration of three-pieces smoother.

In order to ensure the smoother reach the same set-point as the original command, another unit-gain constraint must be added. The integral of the smoother is limited to one to satisfy the requirement. Assuming that p is larger than r, and resulting from constraints (2.23-2.26) and unit-gain constraint, the time-optimal solution of the three-pieces smoother is described by [24]:

$$s_3(\tau) = \begin{cases} \mu_3(e^{-r\zeta_m\omega_m\tau} - e^{-p\zeta_m\omega_m\tau}), 0 \leq \tau \leq (T_1/p) \\ \mu_3 e^{-r\zeta_m\omega_m\tau}(1-\delta_3), (T_1/p) < \tau < (T_1/r) \\ \mu_3(\sigma_3 e^{-p\zeta_m\omega_m\tau} - \delta_3 e^{-r\zeta_m\omega_m\tau}), (T_1/r) \leq \tau \leq (T_1/p + T_1/r) \\ 0, \text{others} \end{cases}, \quad (2.28)$$

where,

$$\delta_3 = e^{2\pi(r/p-1)\zeta_m/\sqrt{1-\zeta_m^2}}, \quad (2.29)$$

$$\sigma_3 = e^{2\pi(p/r-1)\zeta_m/\sqrt{1-\zeta_m^2}}, \quad (2.30)$$

$$\mu_3 = \frac{pr\zeta_m\omega_m}{(p-r)(1-e^{-2\pi\zeta_m/\sqrt{1-\zeta_m^2}})^2}. \quad (2.31)$$

The three-pieces smoother illustrated in Figure 2.8 was derived from the equation (2.28). The curve shows a piecewise continuous profile with three pieces. Convolving the original command with the three-pieces smoother creates a smoothed command. The smoothed command moves the flexible structure inducing tolerable vibrations. Instead of forcing vibrations to zero at the design frequency, the three-pieces smoother limits vibrations to the tolerable level. Meanwhile, the robustness to modeling errors in the

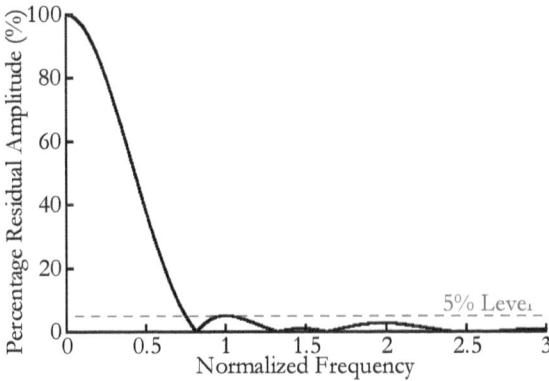

Figure 2.9 Frequency sensitivity for three-pieces smoother.

frequency has been increased by forcing vibrations to zero at two modified frequencies.

The zero-derivative constraint ensures that vibrations could be suppressed below the tolerable level between two modified frequencies. The duration of the three-pieces smoother (2.28) is given by:

$$R_{s3} = (1/p + 1/r) \cdot T_1. \tag{2.32}$$

When the tolerable level, V_{tol}, was set to 5%, numerical solutions for undamped systems are $r=0.7545$ and $p=1.2277$. When the tolerable level was set to zero, amending coefficients for both undamped and damped systems are $p=r=1$. Therefore, the zero tolerable level will create a smoother with specified performance, including zero vibration and derivative at the design frequency and damping ratio.

2.3.2 Sensitivity to Modeling Error

Figure 2.9 shows the frequency sensitivity curve for the three-pieces smoother designed for a zero damping ratio. The 5% insensitivity of the three-pieces smoother ranges from 0.747 to infinity. The three-pieces smoother has more insensitivity at the high frequency, and is less insensitive at the low frequency. As the normalized frequency increases, the magnitude of peaks of the percentage residual amplitude decreases. This low-pass filtering performance will benefit vibration reduction for multi-mode systems. When the three-pieces smoother is designed to reduce vibrations of the first mode, it will also suppress that of higher modes.

2.4 Four-Pieces Smoother

2.4.1 Design of Four-Pieces Smoother

The high mode also have an effect on the system dynamics. Thus, the higher-mode vibrations should be suppressed. In order to create a low-pass filtering effect for the smoother, oscillations at the normalized frequency of two should be suppressed to zero:

$$\int_{\tau=0}^{+\infty} s_4(\tau)e^{2\cdot\omega_m\zeta_m\tau}\sin(2\cdot\omega_m\sqrt{1-\zeta_m^2}\tau)d\tau = 0, \tag{2.33}$$

$$\int_{\tau=0}^{+\infty} s_4(\tau)e^{2\cdot\omega_m\zeta_m\tau}\cos(2\cdot\omega_m\sqrt{1-\zeta_m^2}\tau)d\tau = 0, \tag{2.34}$$

where ζ_m is the design damping ratio, ω_m is the design frequency, and s_4 is the four-pieces smoother.

When the design frequency and damping ratio are correct, the constraints (2.33) and (2.34) would suppress vibrations at the normalized frequency of two, $2\cdot\omega_m$. Nevertheless, real systems often operate with some degree of uncertainty at the high mode. Therefore, another constraint should be added to increase the robustness to changes in the high-mode frequency. The zero derivative of the percent vibration amplitude with respect to frequency at the normalized frequency of two should also be added:

$$\int_{\tau=0}^{+\infty} \tau\cdot s_4(\tau)e^{2\cdot\omega_m\zeta_m\tau}\sin(2\cdot\omega_m\sqrt{1-\zeta_m^2}\tau)d\tau = 0, \tag{2.35}$$

$$\int_{\tau=0}^{+\infty} \tau\cdot s_4(\tau)e^{2\cdot\omega_m\zeta_m\tau}\cos(2\cdot\omega_m\sqrt{1-\zeta_m^2}\tau)d\tau = 0. \tag{2.36}$$

The robustness around the design frequency will maximize by limiting vibrations at two modified frequencies, $p\cdot\omega_m$ and $r\cdot\omega_m$, to zero. Because the design frequency, ω_m, occurs between two modified frequencies, $p\cdot\omega_m$, and, $r\cdot\omega_m$, vibrations at two modified frequencies, $p\cdot\omega_m$, and, $r\cdot\omega_m$, should also be limited to zero:

$$\int_{\tau=0}^{+\infty} s_4(\tau)e^{r\cdot\omega_m\zeta_m\tau}\sin(r\cdot\omega_m\sqrt{1-\zeta_m^2}\tau)d\tau = 0, \quad 0 < r \leq 1, \tag{2.37}$$

$$\int_{\tau=0}^{+\infty} s_4(\tau)e^{r\cdot\omega_m\zeta_m\tau}\cos(r\cdot\omega_m\sqrt{1-\zeta_m^2}\tau)d\tau = 0, \quad 0 < r \leq 1, \tag{2.38}$$

$$\int_{\tau=0}^{+\infty} s_4(\tau)e^{p\cdot\omega_m\zeta_m\tau}\sin(p\cdot\omega_m\sqrt{1-\zeta_m^2}\tau)d\tau = 0, \quad p\geq 1, \tag{2.39}$$

$$\int_{\tau=0}^{+\infty} s_4(\tau)e^{p\cdot\omega_m\zeta_m\tau}\cos(p\cdot\omega_m\sqrt{1-\zeta_m^2}\tau)d\tau = 0, \quad p\geq 1. \tag{2.40}$$

The smoothed command should reach the same set-point as the original command, thus the integral of the smoother should also be limited to one. The constraints (2.33-2.40) and unit-gain constraint yield a time-optimal solution of the four-pieces smoother [25]:

$$s_4(\tau) = \begin{cases} M_4 \cdot \left[(1+\sigma_4)\tau\right] \cdot e^{-2\zeta_m\omega_m\tau}, & 0 \leq \tau \leq 0.5T_1 \\ M_4 \cdot \begin{bmatrix} (1+\sigma_4+\sigma_4 K_4 - K_4)T_1 \\ +(2K_4 - 2K_4\sigma_4 - \sigma_4 - 1)\tau \end{bmatrix} \cdot e^{-2\zeta_m\omega_m\tau}, & 0.5T_1 < \tau \leq T_1 \\ M_4 \cdot \begin{bmatrix} K_4(3 - K_4\sigma_4 - 3\sigma_4 - K_4)T_1 \\ +K_4(K_4 + \sigma_4 K_4 + 2\sigma_4 - 2)\tau \end{bmatrix} \cdot e^{-2\zeta_m\omega_m\tau}, & T_1 < \tau \leq 1.5T_1 \\ M_4 \cdot \left[2K_4^2(1+\sigma_4)T_1 - K_4^2(1+\sigma_4)\tau\right] \cdot e^{-2\zeta_m\omega_m\tau}, & 1.5T_1 < \tau \leq 2T_1 \end{cases}, \tag{2.41}$$

where σ_4 is the coefficient, and the coefficients, K_4 and M_4, are:

$$K_4 = e^{2\pi\zeta_m/\sqrt{1-\zeta_m^2}}, \tag{2.42}$$

$$M_4 = \zeta_m^2\omega_m^2 / (1-K_4^{-1})^2. \tag{2.43}$$

Figure 2.10 shows that the four-pieces smoother (2.41) is a piecewise continuous function with four pieces. Continuous transitions between boundary conditions exist in each of pieces. The Laplace transform of the four-pieces smoother resulting from equation (2.41) is given by [25]:

$$s_4(s) = \frac{M_4 \cdot \begin{pmatrix} (1+\sigma_4) + 2K_4^{-1}(K_4 - \sigma_4 K_4 - 1 - \sigma_4)e^{-0.5T_1\cdot s} \\ +K_4^{-2}(K_4^2 + \sigma_4 K_4^2 - 4K_4 + 4\sigma_4 K_4 + 1 + \sigma_4)e^{-T_1\cdot s} \\ +2K_4^{-2}(-K_4 - \sigma_4 K_4 + 1 - \sigma_4)e^{-1.5T_1\cdot s} + K_4^{-2}(1+\sigma_4)e^{-2T_1\cdot s} \end{pmatrix}}{(s + 2\zeta_m\omega_m)^2}. \tag{2.44}$$

A local extremum in the percent vibration amplitude arises around the design frequency. Let $v\cdot\omega_m$ (satisfies $r\leq v\leq p$) be the frequency at the local extreme value. Then the percent vibration amplitude at the local extremum should be limited to a tolerable level, V_{tol}.

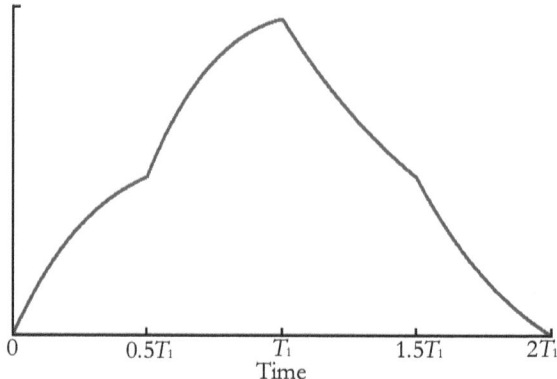

0        $0.5T_1$ T_1 $1.5T_1$ $2T_1$

Time

Figure 2.10 Curve of four-pieces smoother.

$$\left[\sqrt{\begin{array}{c}[\int_{\tau=0}^{+\infty} s_4(\tau)e^{\zeta\omega\tau}\sin(\omega\sqrt{1-\zeta^2}\tau)d\tau]^2 \\ +[\int_{\tau=0}^{+\infty} s_4(\tau)e^{\zeta\omega\tau}\cos(\omega\sqrt{1-\zeta^2}\tau)d\tau]^2\end{array}}\right]_{\substack{\omega=v\cdot\omega_m \\ \zeta=\zeta_m}} = V_{tol}. \tag{2.45}$$

In addition, the derivative of equation (2.45) with respect to frequency should also be limited to zero:

$$\left[\frac{d\left\{\begin{array}{c}[\int_{\tau=0}^{+\infty} s_4(\tau)e^{\zeta\omega\tau}\sin(\omega\sqrt{1-\zeta^2}\tau)d\tau]^2 \\ +[\int_{\tau=0}^{+\infty} s_4(\tau)e^{\zeta\omega\tau}\cos(\omega\sqrt{1-\zeta^2}\tau)d\tau]^2\end{array}\right\}}{d\omega}\right]_{\substack{\omega=v\cdot\omega_m \\ \zeta=\zeta_m}} = 0. \tag{2.46}$$

Solving equations (2.45) and (2.46) yields the coefficient, σ_4. However, the closed-form solution is difficult because of nonlinearity. Instead, substituting equation (2.41) into equations (2.45) and (2.46) yields a numerical solution. The solution of the coefficient, σ_4, is zero, when the tolerable level of the vibrations, V_{tol}, is set to zero.

2.4.2 Sensitivity to Modeling Error

The four-pieces smoother is a function of design frequency and damping ratio. However, the frequency might not be known accurately in a real system, then it is important to estimate how the modeling error in the

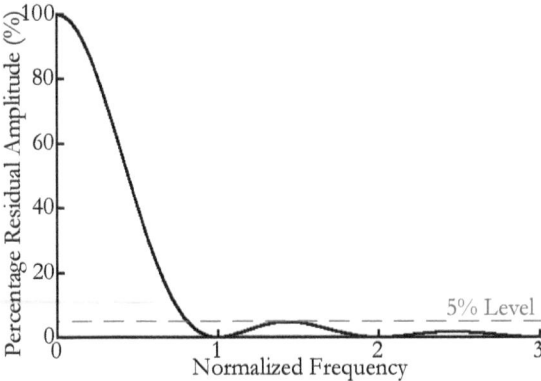

Figure 2.11 Frequency sensitivity curve with four-pieces smoother.

frequency translates into the percent residual amplitude. Figure 2.11 shows the frequency insensitive curve for the four-pieces smoother designed for an undamped system. There are troughs at the design frequency and integer normalized frequencies. The 5 % insensitivity of the four-pieces smoother is from 0.81 to infinity. Meanwhile, damping ratio has little impacts on vibrations with the four-pieces smoother. This conclusion is fortunate because it is generally challenging to estimate the damping ratio accurately under real conditions.

2.5 Smoother for Duffing Oscillators

2.5.1 Single-Mode Duffing Oscillators

The approximate solution to a damped Duffing oscillator resulting from an impulse function, s_{SD}, can be derived by using a perturbation method. Then the first-order approximate solution is:

$$
\begin{aligned}
q(t) = & \int_{\tau=0}^{+\infty} s_{SD}(\tau) \cdot \psi_1 \cdot e^{-\zeta\bar{\omega}(t-\tau)} \sin[\bar{\omega}\sqrt{1-\zeta^2}(t-\tau)+\varphi_1]d\tau \\
& + \int_{\tau=0}^{+\infty} v(\tau) \cdot \psi_2 \cdot e^{-\zeta\bar{\omega}(t-\tau)} \sin[\bar{\omega}\sqrt{1-\zeta^2}(t-\tau)+\varphi_2]d\tau \quad , \\
& + \int_{\tau=0}^{+\infty} v(\tau) \cdot \psi_3 \cdot e^{-3\zeta\bar{\omega}(t-\tau)} \sin[3\bar{\omega}\sqrt{1-\zeta^2}(t-\tau)+\varphi_3]d\tau
\end{aligned}
\tag{2.47}
$$

where ψ_1, ψ_2, and ψ_3 are vibration-contribution functions, φ_1, φ_2 and φ_3 are phase functions, and $\bar{\omega}$ is the nonlinear frequency. The parameter, v, is given by:

$$v(\tau) = \int h(t) \cdot s_{SD}(\tau - t) dt$$
$$h(t) = \int s_{SD}(y) \cdot s_{SD}(t - y) dy \qquad (2.48)$$

The nonlinear frequency, $\bar{\omega}$, is given by:

$$\bar{\omega} = \omega \cdot \sqrt{1 + 0.75e \cdot [q^2 + (\zeta q + \dot{q}/\omega)^2]}, \qquad (2.49)$$

where ω, e, q are the linear frequency, nonlinear stiffness parameter, and amplitude of the Duffing oscillator, respectively.

The nonlinear frequency is dependent on the nonlinear stiffness parameter, e, the linear frequency, ω, the time-dependent vibration amplitude, q, and its derivative, $\dot{q}$. As the linear frequency, ω, and nonlinear stiffness parameter, e, increase, the nonlinear frequency increases. Increasing vibration amplitudes increases the nonlinear frequency in the case of $e>0$ (hardening spring type). Meanwhile, increasing vibration amplitudes decreases the nonlinear frequency in the case of $e<0$ (softening spring type).

The amplitude of approximate solution (2.47) can be written as:

$$A(t) = \psi_1 \cdot e^{-\zeta \bar{\omega} t} \sqrt{[S_1(\zeta, \bar{\omega})]^2 + [C_1(\zeta, \bar{\omega})]^2}$$
$$+ \psi_2 \cdot e^{-\zeta \bar{\omega} t} \sqrt{[S_2(\zeta, \bar{\omega})]^2 + [C_2(\zeta, \bar{\omega})]^2}, \qquad (2.50)$$
$$+ \psi_3 \cdot e^{-3\zeta \bar{\omega} t} \sqrt{[S_3(\zeta, \bar{\omega})]^2 + [C_3(\zeta, \bar{\omega})]^2}$$

where

$$S_1(\zeta, \bar{\omega}) = \int_{\tau=0}^{+\infty} s_{SD}(\tau) \cdot e^{\zeta \bar{\omega} \tau} \sin[\bar{\omega}\sqrt{1-\zeta^2} \cdot \tau] d\tau, \qquad (2.51)$$

$$C_1(\zeta, \bar{\omega}) = \int_{\tau=0}^{+\infty} s_{SD}(\tau) \cdot e^{\zeta \bar{\omega} \tau} \cos[\bar{\omega}\sqrt{1-\zeta^2} \cdot \tau] d\tau, \qquad (2.52)$$

$$S_2(\zeta, \bar{\omega}) = \int_{\tau=0}^{+\infty} v(\tau) \cdot e^{\zeta \bar{\omega} \tau} \sin[\bar{\omega}\sqrt{1-\zeta^2} \cdot \tau] d\tau, \qquad (2.53)$$

$$C_2(\zeta, \bar{\omega}) = \int_{\tau=0}^{+\infty} v(\tau) \cdot e^{\zeta \bar{\omega} \tau} \cos[\bar{\omega}\sqrt{1-\zeta^2} \cdot \tau] d\tau, \qquad (2.54)$$

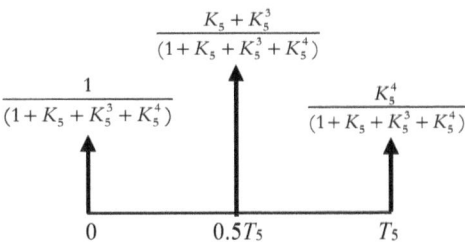

Figure 2.12 Three-impulses SD function.

$$S_3(\zeta,\bar{\omega}) = \int_{\tau=0}^{+\infty} v(\tau) \cdot e^{3\zeta\bar{\omega}\tau} \sin[3\bar{\omega}\sqrt{1-\zeta^2} \cdot \tau]d\tau , \qquad (2.55)$$

$$C_3(\zeta,\bar{\omega}) = \int_{\tau=0}^{+\infty} v(\tau) \cdot e^{3\zeta\bar{\omega}\tau} \cos[3\bar{\omega}\sqrt{1-\zeta^2} \cdot \tau]d\tau . \qquad (2.56)$$

Limiting equations (2.51)-(2.56) to zero yields zero vibrations of the approximate response. The modified command should also reach the same position as the original driving command. Thus, a unity gain constraint should be added:

$$\int s_{SD}(\tau) = 1 . \qquad (2.57)$$

Resulting from equations (2.51)-(2.56) and constraint (2.57), an impulse function for the single-mode Duffing oscillator (SD) is given by [26]:

$$s_{SD}(\tau) = \begin{bmatrix} \dfrac{1}{(1+K_5+K_5^3+K_5^4)} & \dfrac{K_5+K_5^3}{(1+K_5+K_5^3+K_5^4)} & \dfrac{K_5^4}{(1+K_5+K_5^3+K_5^4)} \\ 0 & 0.5T_5 & T_5 \end{bmatrix},$$

$$\qquad (2.58)$$

where

$$K_5 = e^{(-\pi\zeta/\sqrt{1-\zeta^2})}, \qquad (2.59)$$

$$T_5 = \dfrac{2\pi}{\bar{\omega} \cdot \sqrt{1-\zeta^2}} . \qquad (2.60)$$

The three-impulses SD function, as a function of nonlinear frequency, $\bar{\omega}$, and damping ratio, ζ, is shown in Figure 2.12. When the nonlinear frequency and damping ratio are found accurately, the approximate response would be suppressed to zero. Both the nonlinear frequency, $\bar{\omega}$, and the corresponding period, T_5, depend on the vibration amplitude. The duration of the three-impulses SD function is a damped vibration period, T_5. The Laplace transform of the three-impulses SD function can be derived from equation (2.58) [26]:

$$s_{SD}(s) = \frac{1}{(1+K_5+K_5^3+K_5^4)} + \frac{(K_5+K_5^3)\cdot e^{-0.5T_5 s}}{(1+K_5+K_5^3+K_5^4)} + \frac{K_5^4\cdot e^{-T_5 s}}{(1+K_5+K_5^3+K_5^4)}.$$

$$(2.61)$$

2.5.2 Multi-Mode Duffing Oscillators

Vibrations of multi-mode Duffing oscillators can also be suppressed by the three-impulses SD function. The three-impulses SD function for each mode of the uncoupled Duffing oscillators is designed independently, and then combining those together produces a combination function. Vibrations of multi-mode Duffing oscillators can be reduced by the combination function. Nevertheless, it is very challenging to estimate high-mode frequencies under practical engineering conditions. However, the modeling error in the high-mode frequency might be large so that designing a suitable controller is impossible.

The low-pass filter benefits vibration suppression of high-mode vibrations. Then vibrations at the even frequency, $2\bar{\omega}$, should also be zero:

$$\int_{\tau=0}^{+\infty} s_{MD}(\tau)\cdot e^{2\zeta\bar{\omega}\tau}\sin[2\bar{\omega}\sqrt{1-\zeta^2}\cdot\tau]d\tau = 0,$$

$$(2.62)$$

$$\int_{\tau=0}^{+\infty} s_{MD}(\tau)\cdot e^{2\zeta\bar{\omega}\tau}\cos[2\bar{\omega}\sqrt{1-\zeta^2}\cdot\tau]d\tau = 0.$$

$$(2.63)$$

A smoother for multi-mode Duffing oscillators (MD) can be derived by limiting equations (2.51)-(2.56) to zero and solving constraints (2.62)-(2.63) [26]:

$$s_{MD}(\tau) = \begin{cases} M_5\cdot\tau\cdot e^{-2\zeta\bar{\omega}\tau}, & 0\leq\tau\leq 0.5T_5 \\ M_5\cdot\begin{bmatrix}(-0.5K_5^{-1}+1-0.5K_5)T_5 \\ +(K_5^{-1}-1+K_5)\tau\end{bmatrix}\cdot e^{-2\zeta\bar{\omega}\tau}, & 0.5T_5\leq\tau\leq T_5 \\ M_5\cdot\begin{bmatrix}(1.5K_5^{-1}-1+1.5K_5)T_5 \\ +(-K_5^{-1}+1-K_5)\tau\end{bmatrix}\cdot e^{-2\zeta\bar{\omega}\tau}, & T_5\leq\tau\leq 1.5T_5 \\ M_5\cdot[2T_5-\tau]\cdot e^{-2\zeta\bar{\omega}\tau}, & 1.5T_5\leq\tau\leq 2T_5 \end{cases}$$

$$(2.64)$$

where

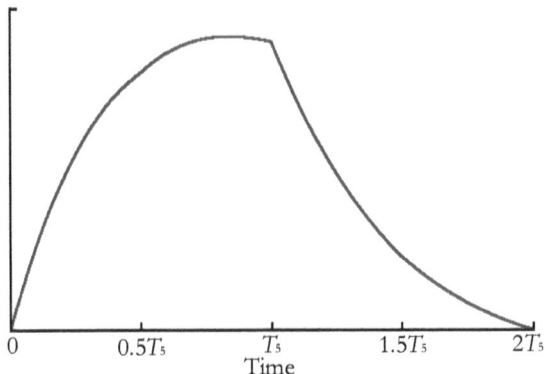

Figure 2.13 Profile of the MD smoother.

$$M_5 = \frac{4\zeta^2 \cdot \bar{\omega}^2}{(1 + K_5 + K_5^3 + K_5^4)(1 - 2K_5^2 + K_5^4)} \cdot \tag{2.65}$$

The profile of the MD smoother (2.64) is shown in Figure 2.13. The MD smoother is a piecewise continuous function involving with nonlinear frequency, $\bar{\omega}$, and damping ratio, ζ. The Laplace transform of the MD smoother resulting from (2.64) is given by [26]:

$$s_{MD}(s) = \frac{M_5}{(s + 2\zeta\bar{\omega})^2} \cdot \begin{bmatrix} 1 + (K_5 - 2K_5^2 + K_5^3) \cdot e^{-0.5T_5 s} \\ +(-2K_5^3 + 2K_5^4 - 2K_5^5) \cdot e^{-T_5 s} \\ +(K_5^5 - 2K_5^6 + K_5^7) \cdot e^{-1.5T_5 s} + M_5 K_5^8 \cdot e^{-2T_5 s} \end{bmatrix} \cdot \tag{2.66}$$

The MD smoother is a combination of multi-notch and low-pass filter, while the SD function is a notch filter. The SD function suppresses vibrations of the first-mode Duffing oscillator, but cannot control that of high-mode Duffing oscillators. The high-mode vibrations might also have some effects on the flexible dynamic system. The MD smoother not only suppresses vibrations of the first-mode Duffing oscillator, and but also attenuates vibrations of high-mode Duffing oscillators. Consequently, the MD smoother does not need to estimate high-mode frequencies. It is generally easier to estimate the first-mode frequency. Thus, this feature exhibits the advantage of control implementation.

The duration of the MD smoother is twice as long as the damped vibration period, T_5. The long duration will slow down the operating speed of flexible structures. Nevertheless, the MD smoother reduces vibrations of flexible structures. Then the flexible structure can move at a higher speed without generating additional vibrations. Additionally, small variations in

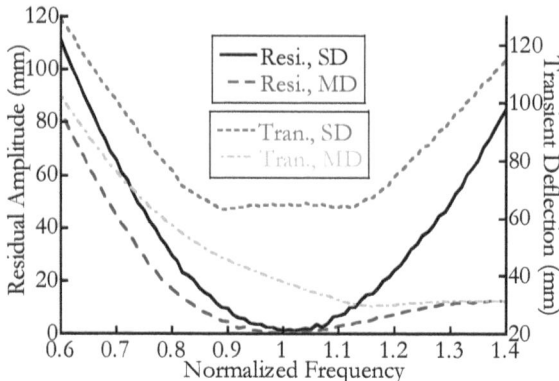

Figure 2.14 Simulated results against modeling errors in the frequency.

the system parameter might cause large changes of the high-mode frequency. Robustness in the high-mode frequency must trade off against an increase of rise time.

2.5.3 Sensitivity to Modeling Error

Both the three-impulse SD function and MD smoother require the use of the nonlinear frequency and damping ratio. However, the nonlinear frequency and damping ratio may not be known accurately. This section uses the dynamics of the single-link flexible manipulator to conduct the numerical verification of the effectiveness and robustness of the three-impulse SD function and MD smoother.

The changes of simulated transient and residual amplitudes with modeling errors in the frequency are shown in Figure 2.14. The Young's modulus, moment of inertia, linear mass density of the beam, beam length, mass ratio, damping ratio, maximum slewing velocity, and maximum slewing acceleration in simulations are 2.06 x 10^5 MPa, 3.449 mm^4, 0.3143 kg/m, 95 cm, 0.5, 0.03, 20°/s, and 200°/s^2, respectively. The ratio of the nonlinear frequency to the designed frequency is defined as the normalized frequency. Increasing modeling errors increases the transient deflection with the three-impulse SD function. When the normalized frequency is about 1.0, the transient deflection is smaller. In contrast, the transient deflection with the MD smoother decreases as the normalized frequency increases. With both the SD function and MD smoother, residual amplitudes are limited to nearly zero around the normalized frequency of 1.0, and show a similar pattern to the corresponding transient deflection. Residual amplitudes with both the SD function and MD smoother increase

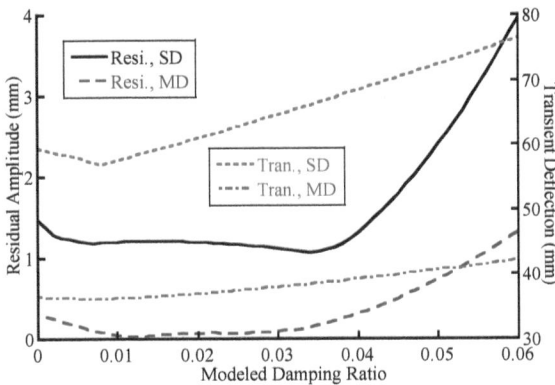

Figure 2.15 Simulated results against modeling errors in the damping.

as the modeling errors in the frequency increase. However, the MD smoother provides better performance in the robustness.

The simulated transient and residual results for various modeling errors in the damping ratio is shown in Fig. 2.15. The real damping ratio is fixed at 0.03. The designed damping ratio ranged from 0 to 0.06. The transient deflection with the three-impulse SD function decreases until the design damping ratio increases to 0.008. Then the transient deflection of the SD function increases beyond this value. The transient deflection with the MD smoother increases slightly with increasing design damping ratio. The residual amplitudes are suppressed to a low level by using both the SD function and MD smoother. However, the residual amplitudes increase quickly for the SD function when the designed damping ratio increases from 0.04.

2.6 Comparison with Other Control Techniques

In order to control multi-mode vibrations for flexible structures, numerous works have been achieved by using the feedback controller and input shaping approaches. However, the difficulty of measuring or estimating high-mode vibrations is an actual obstacle towards the feedback-control application. Meanwhile, the feedback controller cannot control vibrations of the high-mode frequency when the bandwidth of actuators or sensors lies outside of the high-mode frequency. Therefore, feedback-control techniques are challenging to reduce high-mode vibrations for flexible structures.

The input shaper is a series of impulses, which are convolved with the driving command to move the flexible structure. The input shaper is designed for each modes of the flexible structure independently, and then convolving them together for multi-mode systems. However, high-mode frequencies under real conditions are difficult to measure. Convolving an infinite number of input shapers for infinite mode systems is impossible. Therefore, both the feedback control and input shaper are directed at controlling oscillations of the first few modes.

The command smoothing techniques limit vibrations of an infinite number of modes for flexible structures. The one-piece smoother, two-pieces smoother, three-pieces smoother, four-pieces smoother, and MD smoother need to estimate natural frequency and damping ratio. The smoother is designed to suppress first-mode oscillations of the flexible structure, and then it will also control high-mode vibrations. Hence, the design of the smoother does not need to estimate high-mode frequencies. This conclusion is lucky because estimation of the low frequency is generally easier than that of the high frequency.

The smoother can be designed by using the first-mode frequency and damping ratio. Then the smoother could reduce vibrations of total modes by using the first-mode frequency. Nonetheless, the first mode has the longest vibration period. This is a drawback because the long duration of the smoother will slow down the flexible structure. Because the smoother benefits vibration suppression, the maximum speed of the flexible structure might increase. The high-speed command is modified by the smoother to create a smoothed command. The smoothed command would drive the flexible structure faster.

2.7 Chapter Summary

This chapter described five types of command smoother. The one-piece smoother, two-pieces smoother, three-pieces smoother, four-pieces smoother, and smoother for multi-mode Duffing oscillators are time-optimal solutions resulting from various vibrational constraints, and are robust to modeling errors in the frequency. The presented smoothers can suppress vibrations of an infinite number of modes in flexible structures. The higher-order derivative smoother has also been investigated. While taking additional derivatives will extend the insensitivity, it incurs a rise time penalty.

Chapter 3. Application in Bridge Cranes

Industrial bridge cranes offer material-handling services throughout the world. However, flexible structures degrade their effectiveness and safety. Payload oscillations caused by both intentional motion commanded by the human operator and by external disturbances such as wind are major limitations. Therefore, it is critical to reduce effectively payload oscillations by using presented smoothers in the Chapter 2.

3.1 Single-Pendulum Cranes

Numerous scientists have focused on providing solutions to the payload oscillation problem posed by the industrial crane [27]. The work can roughly be broken into two categories: feedback controller and input shaping. The feedback control strategies use measurements of the payload swing to reduce vibrations. Masoud presented a feedback controller to suppress oscillations of the rotary crane. A damping effect in the crane system was created by careful selection of the gain and time delay of the controller. The control performance was verified on rotary cranes [28], ship-mounted telescopic cranes [29], and quay-side container cranes [30]. Solihin designed a fuzzy-tuned proportional-integral-derivative controller to suppress vibrations of the gantry crane. The controller is robustness to changing parameters [31].

Input shaping techniques can effectively control the payload swing of many types of industrial cranes including bridge cranes [32-33], tower cranes [34-35], boom cranes [36-37], and container cranes [38]. The first input shaper is called zero vibration (ZV) shaper, which limits vibrations to zero at the design frequency and damping ratio. Since then, robust input shapers have been produced, such as zero vibration and derivative (ZVD) shaper, extra-insensitive (EI) shaper, and specified-insensitive (SI) shaper. The fundamental compromise is that the increased robustness provided by those advanced shapers comes at the cost of additional rise time.

3.1.1 Oscillation Reduction by Two-Pieces Smoother

Experiments were performed on a planar bridge crane shown in Figure 3.1 to verify some of key results. A Panasonic AC servomotor with encoder drove the trolley. A personal computer was used for program development and user interface. A DSP-based motion control card (Googol GT-400-SV-PCI) connected the personal computer to a servo amplifier. The original command is sent to the smoother algorithm, and then generates the smoothed command for the drive.

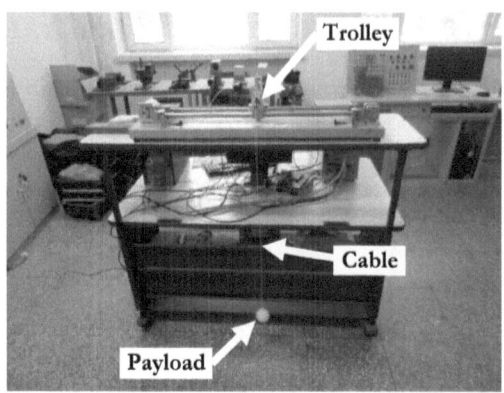

Figure 3.1 A planar single pendulum bridge crane.

A planar bridge crane in Figure 3.1 is modeled as a single pendulum on a trolley. Outputs are the trolley position and the swing angle of the payload. The trolley can be modeled as a mass applied an actuator force. A tennis ball was served as the payload. A weightless cable suspended the ball to the trolley. It is assumed that the payload swing does not affect trolley motions. This is because a large mechanical impedance exists in the drive system, and the swing angle is a small factor. The suspension cable length is also assumed to keep constant during the trolley motion.

The natural frequency of single-pendulum oscillation is given by:

$$\omega = \sqrt{\frac{g}{l}}. \tag{3.1}$$

However, the suspension cable length cannot be measured accurately. Then it is important to investigate vibrations for various estimation errors in the suspension cable length.

A series of experiments were performed by using the two-pieces smoother to move the single pendulum crane when the modeled suspension length was set to 100 cm (corresponding to natural frequency of 0.50 Hz). Figure 3.2 shows experimental and simulated results of residual vibrations as a function of the suspension length. The suspension cable length varied from 40 cm (corresponding to natural frequency of 0.79 Hz) to 160 cm (corresponding to natural frequency of 0.39 Hz). The experimental results follow the general shape as the simulated curve. The two-pieces smoother suppressed the payload swing near the design frequency of 0.50 Hz. The residual amplitude with the two-pieces smoother increased sharply as the suspension length increased from 100 cm. This is

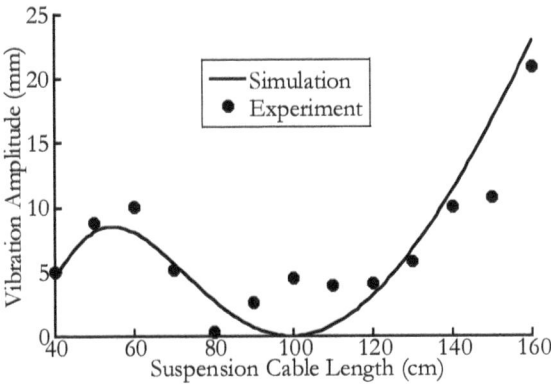

Figure 3.2 Residual vibrations as a function of suspension length.

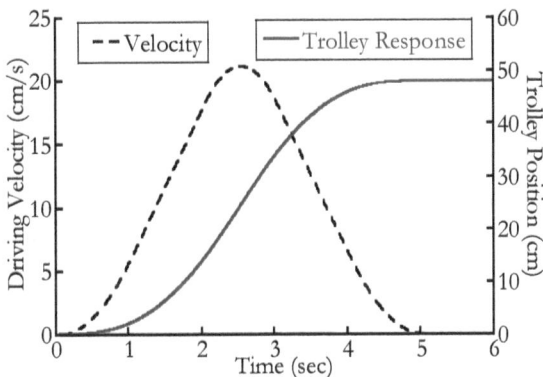

Figure 3.3 Trolley response to smoothed commands.

because more insensitivity at higher frequencies and less insensitivity at low frequencies for the smoother.

The effect when the two-pieces smoother was used to move the trolley 48 cm is shown in Figure 3.3. A trapezoidal velocity profile was used as the original command. The smoothed velocity command was an s-curve profile, which had smooth transitions between boundary conditions.

3.1.2 Oscillation Control by Three-Pieces Smoother

Experiments were performed on a planar bridge crane to verify some of key results of the three-pieces smoother. The residual amplitude to a set trapezoidal velocity profile for driving distance between 5 cm and 55 cm is shown in Figure 3.4. Without the controller, large oscillations were caused on the planar bridge crane when the trolley was driven by a trapezoidal

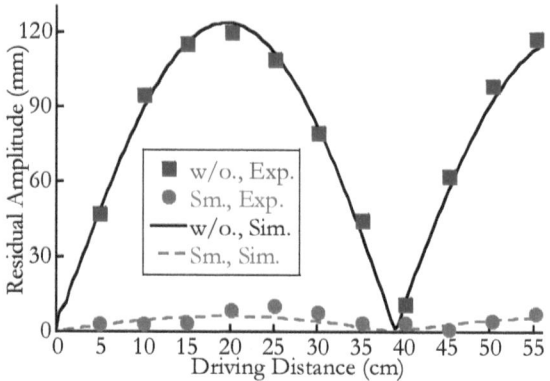

Figure 3.4 Residual vibrations induced by driving motions.

velocity command. Peaks and troughs in the residual amplitude occur with varying driving distances. When the trolley starts, the acceleration will cause large payload swing. Moreover, when the trolley stops, it will also induce additional swing of the payload. Those two oscillations are sometimes out of phase, thereby causing troughs. Sometimes oscillations are in phase, thereby inducing peaks. Figure 3.4 also shows the residual amplitude to the smoothed command. Most of residual vibrations were reduced by the three-pieces smoother. These experiments verified that the three-pieces smoother can also effectively control the payload swing of the single-pendulum crane.

A series of experiments were performed to move planar cranes by using the three-pieces smoother. Figure 3.5 shows residual amplitudes as a function of the suspension cable length. The suspension cable length varies when the modeled suspension length is set to 1 m. The three-pieces smoother suppressed most of oscillations between 10 cm and 170 cm. The residual amplitudes will increase faster when the suspension length increases from 170 cm. This is because the three-pieces smoother provides more insensitivity at higher frequencies and less insensitivity at low frequencies. The experimental findings reported here confirm that the three-pieces smoother can control payload swing of industrial cranes. Furthermore, research findings proved that the three-pieces smoother is more insensitive at higher frequencies.

3.2 Double-Pendulum Cranes

Skilled operators reduce much of oscillations of single-pendulum crane manually by causing deceleration oscillations, which cancel oscillations

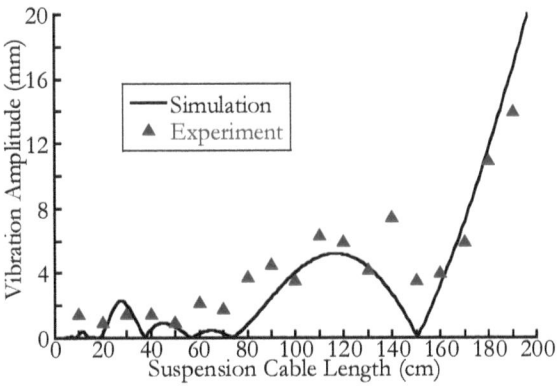

Figure 3.5 Residual vibrations from three-pieces smoother.

induced during the acceleration [39]. However, the double-pendulum dynamics exhibit in certain types of payloads and hoisting mechanisms. Skilled operators are very challenging to eliminate oscillations manually in this case. Therefore, swing-reduction controllers should be equipped in double-pendulum cranes to achieve safe and efficient operations.

Many scientists have worked to provide solutions to the challenging problem posed by the double-pendulum dynamics. The work can roughly be broken into two categories: feedback control and input shaping. Feedback control strategies use measurement and estimation of system state to suppress vibrations [40-46]. However, difficulty of accurately sensing the swing, and the conflict between the computer-based controller and actions of human operator are obstacles toward the application of the feedback controller [47]. Input-shaping technique can effectively reduce payload oscillations of many types of double-pendulum cranes including bridge cranes [48-51], tower cranes [52-53], boom cranes [54], and multi-hoist cranes [55]. While the input shaping technique reduces oscillations induced by intentional motions commanded by the human operator. It cannot reject external disturbances [56].

A schematic representation of a planar double-pendulum bridge crane is shown in Figure 3.6. A force applied to the trolley for moving the crane. A hook of mass, m_h, is attached to the trolley by using the suspension cable of length, l_h. The rigging cable of length, l_p, hangs below the hook and supports a payload. The payload is modeled as a point mass, m_p. It is assumed that both the suspension cable length and rigging cable length do not change during the operation. Then linearized equations of motions are given by:

$$\ddot{\alpha}_h = -\frac{g}{l_h}\alpha_h + \frac{gR}{l_h}\alpha_p - \frac{a}{l_h},$$

(3.2)

29

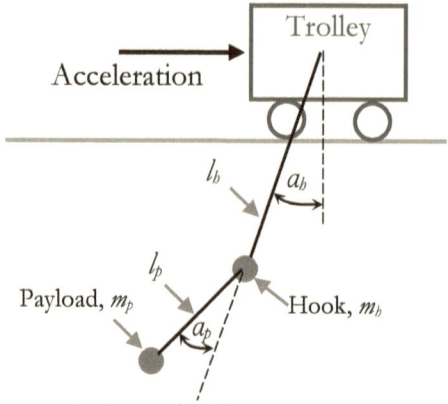

Figure 3.6 A planar double-pendulum bridge crane.

$$\ddot{\alpha}_p = \frac{g}{l_b}\alpha_b - (\frac{g}{l_p} + \frac{gR}{l_p} + \frac{gR}{l_b})\alpha_p + \frac{a}{l_b},$$
(3.3)

where a is the acceleration of the trolley, R is the ratio of the payload mass to the hook mass, and a_b and a_p are swing angle of the suspension cable and rigging cable, respectively. Then, natural frequencies of the planar double-pendulum bridge crane are given by:

$$\omega_{2,1} = \sqrt{\frac{g}{2}}\sqrt{(1+R)(\frac{1}{l_b} + \frac{1}{l_p}) \pm \beta},$$
(3.4)

where

$$\beta = \sqrt{(1+R)^2(\frac{1}{l_b} + \frac{1}{l_p})^2 - 4(\frac{1+R}{l_b l_p})}.$$
(3.5)

The natural frequencies (3.4) are dependent on the suspension cable length, rigging cable length, and mass ratio. The first- and second-mode frequencies as a function of payload mass ranged from 50 g to 300 g are shown in Figure 3.7 when the suspension length, rigging length and hook mass were fixed at 60 cm, 60 cm and 59 g, respectively. The frequency of the first mode ranged from 0.47 Hz to 0.50 Hz, while that of the second mode varied from 1.13 Hz to 2.20 Hz. Such information can be used to design the smoother.

The smoother benefits vibration reduction for double-pendulum cranes. This is because the smoother designed by using the first-mode frequency

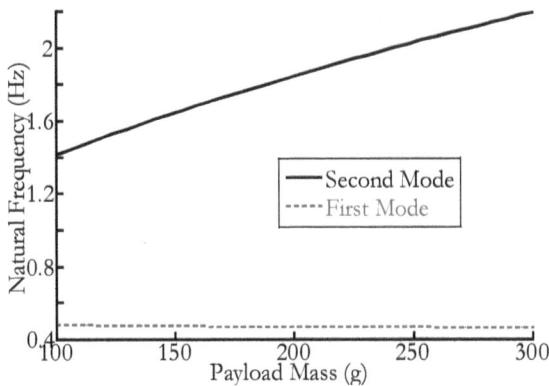

Figure 3.7 Variation of first and second mode frequencies.

will also eliminate oscillations at the second-mode frequency. Thus the second-mode frequency does not need to be estimated. Of course, the first-mode frequency is necessary for the smoother. The damping is assumed to be zero for the double-pendulum crane. When the modeled payload mass is 59 g, a two-pieces smoother for the double-pendulum crane is given by:

$$s_2(s) = \frac{(1 - 2e^{-2.03s} + e^{-4.06s})}{4.1221s^2}. \tag{3.6}$$

A series of experiments were conducted to drive a double-pendulum crane with various mass of the payload. Note that the modeled payload mass was set to 59 g, and the corresponding duration of the two-pieces smoother was 4.06 s. In experiment, the suspension length, rigging length and hook mass were fixed at 60 cm, 60 cm and 59 g, respectively. Figure 3.8 shows the residual vibration amplitude of the hook, while that of the payload is shown in Figure 3.9.

The experimental results follow the general shape as the simulated curve. The experimental data were worse than the simulated curve near the modeling payload mass of 59 g (corresponding to the natural frequencies of 0.49 Hz and 1.19 Hz). This is because the complex double-pendulum crane was simplified to a linearized model.

In the case of the payload mass of 300 g, the experimental results shown in Figure 3.9 were better than the simulated curve because the model was undamped, while the actual system had some small amount of damping. The small damping decreases the sensitivity to modeling error. The two-pieces smoother provided a remarkable suppression in payload swing for all values of hook and payload mass as shown in Figure 3.8 and Figure 3.9.

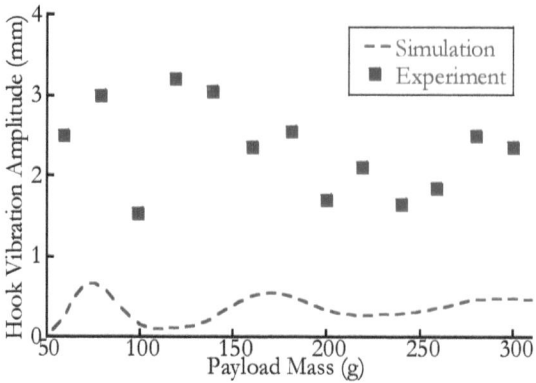

Figure 3.8 Hook residual vibrations.

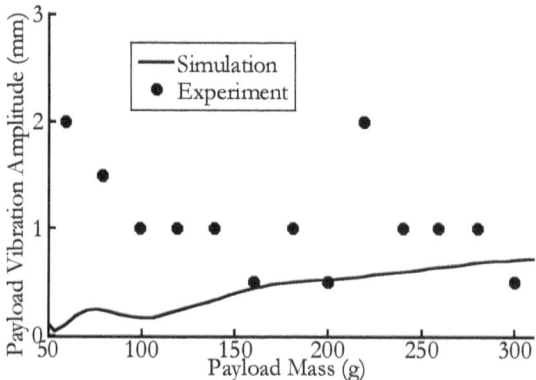

Figure 3.9 Payload residual vibrations.

3.3 Distributed-Mass Payload Cranes

Large objects are suspended from the hook by four rigging cables for transporting bulky loads. This configuration of hoisting mechanism can be seen at shipyards and warehouses. In this case, the load can be modeled as distributed-mass payloads. The distributed-mass payload dynamics are more complicated than typically found on point-mass payload dynamics. Difficulty of moving distributed-mass payloads is due to the payload swing towards the driving direction and the payload twisting about rigging cables. Therefore, the challenging factor for controlling payload oscillations is that the payload twisting must be suppressed. However, few attention has been focused on the control of the payload twisting for industrial cranes.

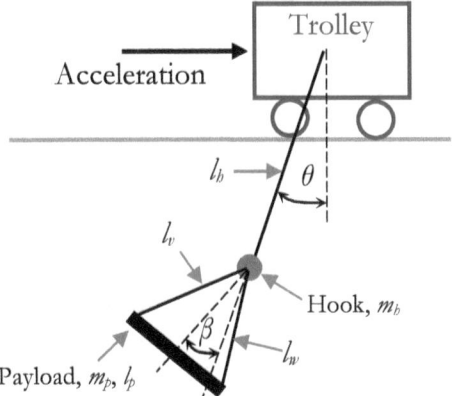

Figure 3.10 Model of double-pendulum cranes with distributed-mass beams.

Therefore, dynamics and control of an actual configuration of bridge crane transporting distributed-mass payloads are essential.

3.3.1 Planar Dynamics

A schematic representation of an actual bridge crane transporting a distributed-mass beam is shown in Figure 3.10. A hook of mass, m_h, is attached to the trolley by a massless suspension cable of length, l_h. Two massless rigging cables of length, l_v and l_w, hang below the hook and support a uniformly distributed-mass beam of mass, m_p, and length, l_p. The input to the model is the trolley acceleration, a. Outputs are the swing angle of the rigging cable relative to the suspension cable, β, and swing angle of the suspension cable, θ. The motion of the trolley is assumed to be unaffected by payload motions due to the large mechanical impedance in the drive system. The suspension and rigging cable lengths are also assumed to be unchanged during the motion. The damping ratio is assumed to be approximately zero. Using the Kane's method, nonlinear equations of the motion are derived [57]:

$$
\begin{aligned}
&R[l_l^2 + l_y(l_y + l_h \cos \beta)]\ddot{\beta} + [Rl_l^2 + l_h^2 + R(l_y^2 + l_h^2 + 2l_y l_h \cos \beta)]\ddot{\theta} \\
&- Rl_y l_h \sin \beta \cdot \dot{\beta}^2 - 2Rl_y l_h \dot{\beta}\dot{\theta}\sin \beta + (gl_h + Rgl_h)\sin \theta \\
&+ Rgl_y \sin(\theta + \beta) = -[l_h \cos \theta + Rl_h \cos \theta + Rl_y \cos(\theta + \beta)]a(t)
\end{aligned}
\qquad (3.7)
$$

33

$$[l_l^2 + l_y^2]\ddot{\beta} + [l_l^2 + l_y(l_y + l_b\cos\beta)]\ddot{\theta} + l_yl_b\sin\beta\cdot\dot{\theta}^2 ,$$
$$+ gl_y\sin(\theta + \beta) = -[l_y\cos(\theta + \beta)]a(t)$$

(3.8)

where,

$$l_l = l_p / (2\sqrt{3}),$$

(3.9)

$$l_y = \frac{1}{2}\sqrt{2l_v^2 + 2l_w^2 - l_p^2} .$$

(3.10)

, R is the ratio of the payload mass to the hook mass, and g is the gravitational constant. It is assumed that the payload swing is small around the equilibrium point, linearized frequencies of this system (3.7) and (3.8) are:

$$\omega_{2,1}^2 = \frac{g(R+1)}{2l_b}(u \pm v) ,$$

(3.11)

where

$$u = \frac{l_y^2 + l_bl_y + l_l^2}{l_y^2 + (R+1)l_l^2} ,$$

(3.12)

$$v = \sqrt{u^2 - \frac{4l_bl_y}{(R+1)(l_y^2 + (R+1)l_l^2)}} .$$

(3.13)

Natural frequencies are dependent on the mass ratio, suspension length, rigging length, and payload length from equation (3.11). Small changes in the system parameter result in large variations in the second-mode frequency. The first-mode frequency changes slightly and the second-mode frequency varies sharply. Therefore, the smoother should have more robustness to changes in the second-mode frequency.

Experiments were performed on a double-pendulum bridge crane moving a distributed-mass beam. A slender beam was attached to hook using two rigging cables, and was served as the distributed-mass payload. A CMOS camera was mounted to the trolley to record horizontal displacements of the hook and two red markers on the beam. Averaging the displacement of two red markers yields the position of the beam centroid.

Experimental responses of the hook and payload for a driving distance of 50 cm are shown in Figure 3.11. The hook mass, payload mass, suspension length, first rigging length, second rigging length, and payload length were held constant at 60 g, and 100 g, 70 cm, 30 cm, 30 cm, and 50

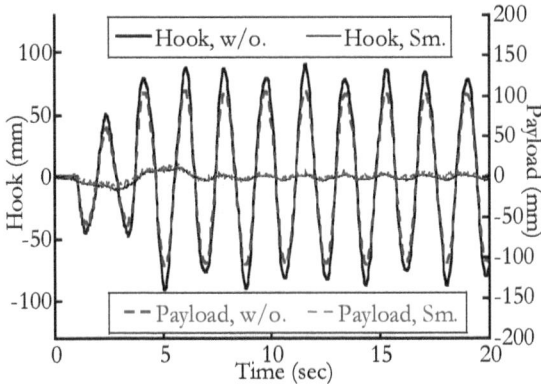

Figure 3.11 Experimental responses for 50 cm driving distance.

cm, respectively. Without the controller, the transient deflection was smaller than the residual amplitude because oscillations caused by the acceleration and deceleration were in phase. The experimental transient deflection and residual amplitude of the payload were 61 mm and 109 mm, respectively. With the two-pieces smoother, the transient deflection and residual amplitude of the payload were 10 mm and 5 mm, respectively. Therefore, the two-pieces smoother dramatically suppresses oscillations of both the hook and payload.

Two sets of experiments were performed to verify the effectiveness of the smoother on suppressing oscillations for variations of system parameters and operation conditions. The first set of experiments examined at the effect of variation in the suspension cable length. The transient and residual amplitudes resulting from these tests are shown in Figure 3.12. The hook mass, payload mass, first rigging length, second rigging length, and payload length were set to be 60 g, 100 g, 30 cm, 30 cm, and 50 cm, respectively. Without the controller, the transient deflection increased with increasing suspension length. A peak in the residual amplitude occurs near the suspension cable length of 65 cm. The experimental results follow the same general shape as the simulated curve. The two-pieces smoother was designed for the modeled suspension length of 90 cm. The two-pieces smoother reduced the transient deflection and residual amplitude by an average of 79.5% and 96.1%, respectively. Therefore, the smoother was robust to changes in the suspension length.

The second set of experiments investigated the effect of variation in the driving distance. The transient deflection and residual amplitude activated for varying driving distances are shown in Figure 3.13. The hook mass, payload mass, suspension length, first rigging length, second rigging length,

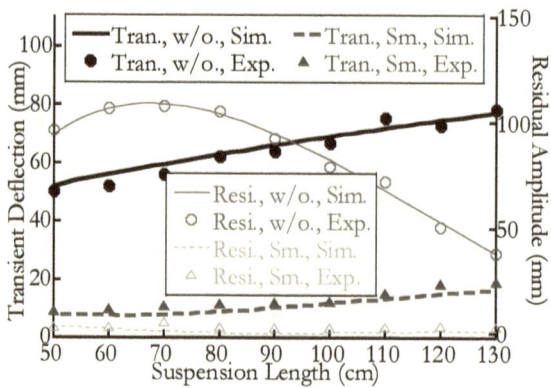

Figure 3.12 Transient and residual vibrations for varying suspension length.

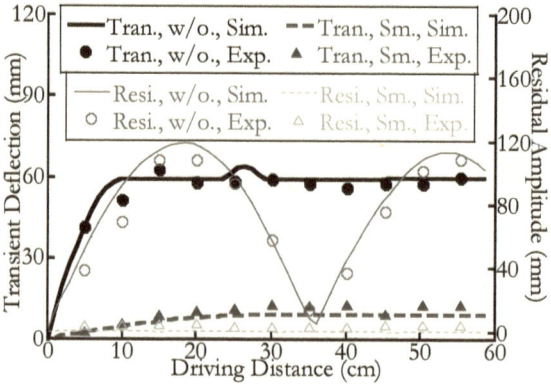

Figure 3.13 Transient and residual vibrations induced by driving distance.

and payload length were set to be 60 g, 100 g, 70 cm, 30 cm, 30 cm, and 50 cm, respectively. Without the controller, the experimental transient deflection would increase with increasing the driving distance before 10 cm. After this point, the transient deflection would depend on the interference between the swing caused by the acceleration and deceleration. Once the oscillation amplitude induced by the interference was larger than the swing amplitude caused by the acceleration, a bump would occur in the transient deflection. Peaks and troughs contain in the experimental residual amplitude as the distance varies because the oscillations caused by the acceleration and deceleration are sometimes in phase or sometimes out of phase. The two-pieces smoother reduced the transient deflection and residual amplitude by an average of 81.8% and 94.8%, respectively. Those

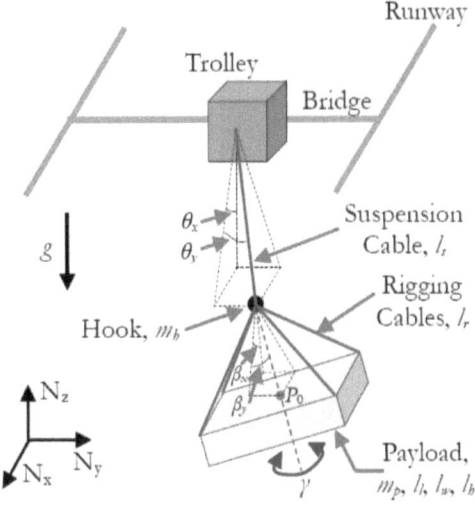

Figure 3.14 A bridge crane carrying distributed-mass payload.

experiments verified that the smoother can effectively control payload oscillations.

3.3.2 Three-Dimensional Dynamics

A. Model

A schematic representation of a bridge crane moving distributed-mass payload is shown in Figure 3.14. The trolley moves along the bridge in the y direction, while the bridge slides along the runway in the x direction. A hook of mass, m_h, is attached to the trolley by a massless suspension cable of length, l_s. Four massless rigging cables of length, l_r, hang below the hook and support a uniformly distributed-mass payload of mass, m_p, length, l_l, width, l_w, and height, l_h. The acceleration of the bridge, a_x, and the acceleration of the trolley, a_y, are inputs to the model. The swing angles of the suspension cable, θ_x and θ_y, and the payload swing angles relative to the suspension cable, β_x and β_y, and the payload twist angle, γ, are outputs. The model assumes that the trolley is significantly more massive than the hook and payload. The suspension and rigging cables are assumed as massless and rigid. The hook is modeled as a point mass. Equations of the motion for this model can be derived using the dynamics software package, MotionGenesis.

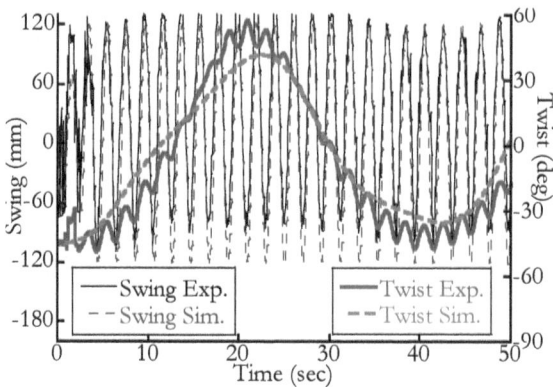

Figure 3.15 Experimental responses for payload swing and twisting.

Experimental and simulated responses to a set trapezoidal velocity command for driving distance of 55 cm are shown in Figure 3.15. The suspension length, payload length and payload mass were fixed at 80 cm, 15 cm and 320 g, respectively. The experimental data follow the same general shape as the simulated curve. The residual amplitude of the payload swing is larger than the transient deflection because the payload swing induced by the deceleration is in phase with that caused by the acceleration. Crane operation will be very challenging when the payload swing and twisting occur. The small-amplitude high-frequency oscillation and the large-amplitude low-frequency oscillation arise in the payload twisting. The small-amplitude high-frequency oscillation of the payload twisting is the only difference between simulated and experimental curves because cables are modeled as massless and rigid, while the actual system has flexible nature of physical structures [58].

B. Dynamic Analysis

The payload twisting cannot be excited by accelerations of the bridge and trolley in the case of zero initial angle of the payload twisting. In this specified condition, the model simplifies the dynamic analysis and provides simple estimates of the frequency of the payload swing. Then linearized equations of the motion relating swing angles, θ_x and β_x, to the acceleration of the bridge are given by:

$$(l_p^2 + l_q^2 + l_s l_q)\ddot{\theta}_x + (l_p^2 + l_q^2)\ddot{\beta}_x + g l_q \theta_x + g l_q \beta_x + l_q a_x = 0, \qquad (3.14)$$

$$(l_s^2 + Rl_p^2 + Rl_s^2 + Rl_q^2 + 2Rl_sl_q)\ddot{\theta}_x + (Rl_p^2 + Rl_q^2 + Rl_ql_s)\ddot{\beta}_x$$
$$+g(l_s + Rl_s + Rl_q)\theta_x + gRl_q\beta_x + (Rl_s + Rl_q + l_s)a_x = 0$$
(3.15)

where

$$l_q = \sqrt{l_r^2 - 0.25l_l^2 - 0.25l_w^2} + 0.5l_b, \tag{3.16}$$

l_p is the radius of gyration of the payload about the N_y axis through its center of mass P_0, g is the gravitational constant, and R is the ratio of the payload mass to the hook mass. Linearized frequencies of the payload swing dynamics modeled in (3.14) and (3.15) are given by:

$$\omega_{2,1} = \sqrt{\frac{g(1+R)}{2l_s}(u \pm v)}, \tag{3.17}$$

where

$$u = \frac{l_q^2 + l_sl_q + l_p^2}{l_q^2 + (R+1)l_p^2}, \tag{3.18}$$

$$v = \sqrt{u^2 - \frac{4l_sl_q}{(R+1)(l_q^2 + l_p^2 + Rl_p^2)}}. \tag{3.19}$$

The first- and second-mode frequencies (3.17) are dependent on the suspension cable length, rigging cables length, mass ratio, and payload size. The second-mode frequency varies more sharply than the first-mode frequency. As a result, the smoother should provide more robustness to variations in the high-mode frequency.

The twisting dynamics of the payload exhibits complicated behavior. The payload swing has a large impact on the frequency of the payload twisting. Increasing the magnitude of the payload swing increases the frequency of the payload twisting. Moreover, the payload size also has some effects on the frequency of the payload twisting. The frequency of the payload twisting is zero when the payload length and width are equal. The frequency of the payload twisting increases with increasing the ratio of the payload length to the payload width.

C. Numerical Verifications

Figure 3.16 shows transient and residual amplitudes of the payload swing induced by various suspension length. The payload length and mass were fixed at 15 cm and 320 g, respectively. Without the controller, the transient deflection increases with increasing suspension length. A local maximum in

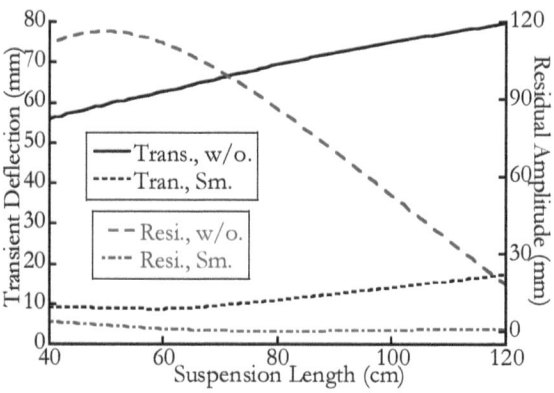

Figure 3.16 Swing amplitude against suspension cable length.

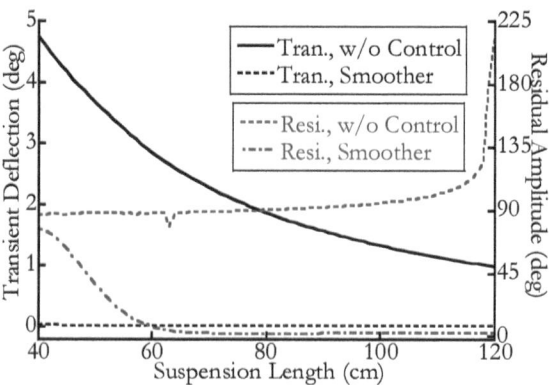

Figure 3.17 Twisting amplitude against suspension cable length.

the residual amplitude arises for a suspension length of 50 cm because oscillations caused by the acceleration and deceleration were in phase. The two-pieces smoother was designed when the modeled suspension length was 80 cm. The two-pieces smoother reduced the transient deflection and residual amplitude by an average of 82.2 % and 98.9 %, respectively.

Figure 3.17 shows transient and residual amplitudes of the payload twisting induced by various suspension length. Without the controller, the transient deflection decreases as the suspension length increases. When the residual amplitude of the payload swing is near its minimum shown in Figure 3.16, a peak in residual amplitude of the payload twisting occurs. This effect can be physically interpreted as the interference between oscillations caused by the acceleration and deceleration. When oscillations caused by the deceleration cancels out oscillations induced by the acceleration, the payload is rotated only by the inertia force caused by transient oscillations of the payload swing. Thus, the spin in one direction causes a sharp increase in the payload twisting. The two-pieces smoother was designed when the modeled suspension length was 80 cm. As the suspension length increases from 40 cm to 60 cm, the controlled residual amplitude of the payload twisting decreases because the payload twisting depends on the size of the payload swing and the duration of the simulation. The two-pieces smoother reduced the transient deflection and residual amplitude by an average of 98.7 % and 89.7 %, respectively.

The transient deflection and residual amplitude of the payload swing as a function of payload length are shown in Figure 3.18. The suspension length was set to 80 cm. Without the controller, the payload length does not have a large effect on the transient deflection and residual amplitude. The two-pieces smoother was designed when the modeled payload length was set to 15 cm. The two-pieces smoother reduced the transient deflection and residual amplitude by an average of 84.1 % and 99.9 %, respectively.

The transient deflection and residual amplitude of the payload twisting for various payload lengths are shown in Figure 3.19. Without the controller, when the payload length is set to 7.5 cm, both the transient deflection and the residual amplitude are limited to zero because the payload length and width are equal. After this point, the transient deflection increases as the payload length increases before 17 cm, and then the payload length has small effect on the transient deflection. Meanwhile, the payload size does not have large effects on the residual amplitude of the payload twisting. The two-pieces smoother was designed when the modeled payload length was set to 15 cm. The two-pieces smoother reduced the transient deflection and residual amplitude by an average of 98.2 % and 97.2 %, respectively. Figures 3.16-3.19 demonstrated that the smoother can robustly reduce oscillations for a wide range of system parameters.

D. Experimental Verifications

Experiments were performed on a bridge crane carrying distributed-mass crates shown in Figure 3.20. A Panasonic AC servomotor with encoder drove the trolley. A digital signal processor (DSP)-based motion control card (Googol GT-400-SV-PCI) connects a personal computer to a servo amplifier. The original command (trapezoidal-velocity profile) is sent to a MATLAB script in the personal computer, which applies the command smoothing algorithm. A crate was attached to hook using four rigging cables. The payload size is 150 mm x 75 mm x 10 mm, and the payload mass is 320 g. The cables are made of Dyneema super braid fishing line. The displacements of two red markers on the crate were measured by a CMOS camera, which was mounted on the trolley. Note that two red

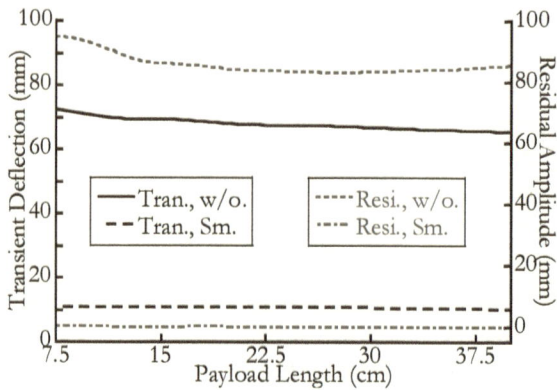

Figure 3.18 Vibration amplitude of payload swing against payload length.

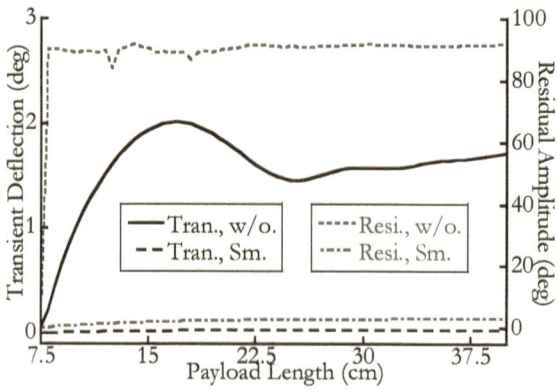

Figure 3.19 Vibration amplitude of payload twisting against payload length.

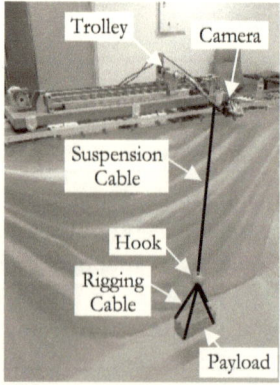

Figure 3.20 Bridge crane transporting distributed-mass crates.

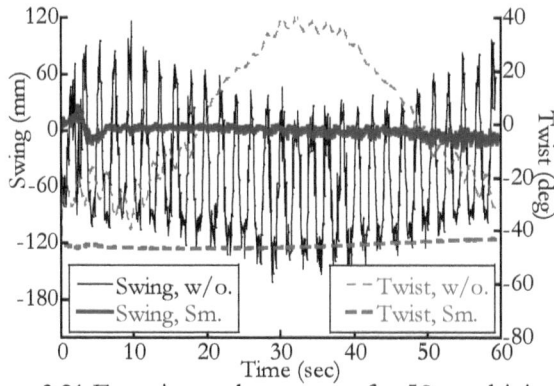
Figure 3.21 Experimental responses for 50 cm driving.

markers were used to calculate the payload swing displacement and the payload twist angle.

Experimental responses of the payload swing and twisting are shown in Figure 3.21. Without the controller, the complex distributed-mass payload dynamics behavior is clearly visible. The two-pieces smoother suppressed oscillations of the payload swing and twisting by an average of 93.4 % and 97.8 %, respectively. Moreover, the two-pieces smoother reduced both the transient deflection and residual amplitude of the payload swing and twisting. Therefore, the smoother was effective in this case.

One set of experiments were performed to verify the effectiveness of the smoother on suppressing oscillations for variations of system parameters and operation conditions. Variations in the suspension length were selected to exam a wide range of possible dynamics. The residual amplitude of the payload swing and twisting for varying suspension lengths is shown in Figure 3.22 and Figure 3.23, respectively. Without the controller, a peak in the payload swing would occur for the suspension length of 50 cm. The residual amplitude of the payload twisting were approximately 90° for most of suspension length except 120 cm. The experimental data were better than the simulated curve at the suspension length of 120 cm. This is because the model was undamped, while the actual system had some small amount of damping. The damping prevented the inertia force from spinning the payload when oscillations caused by deceleration cancel out oscillations induced by the acceleration. The two-pieces smoother was designed when the modeled suspension length was set to 80 cm. The smoother suppressed experimental residual amplitudes of the payload swing and twisting by an average of 95.2 % and 84.9 %, respectively. Those experiments proved that the smoother can effectively suppress oscillations of the payload swing and

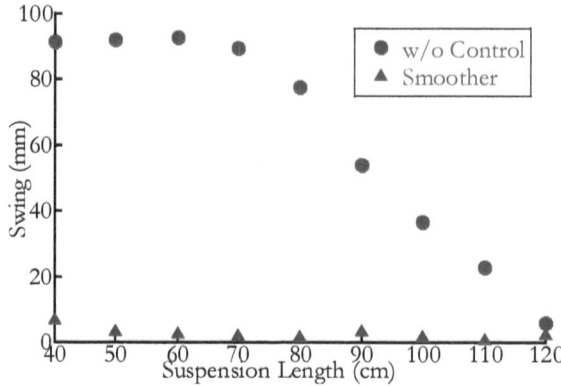

Figure 3.22 Experimental residual amplitude of payload swing for varying suspension cable lengths.

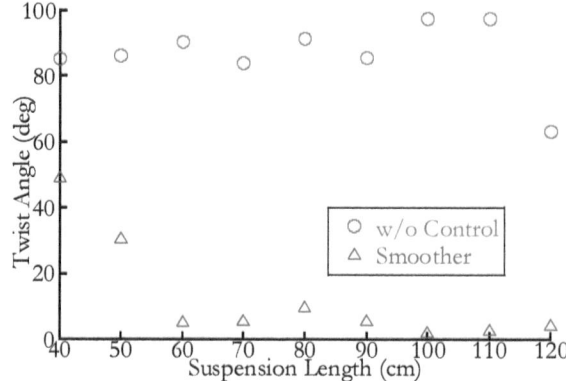

Figure 3.23 Experimental residual amplitude of payload twisting for varying suspension cable lengths.

twisting induced by various system parameters as it was predicted by simulations.

3.4 Wind Rejection

Many scientists designed feedback controller to damp the payload swing of the crane. However, difficulty of accurately sensing the payload and its velocity is a drawback toward the application of the feedback controller. Additionally, the conflict between the computer-based feedback controller and actions of human operator is also an obstacle. Open-loop control schemes for reducing crane payload oscillations include inverse kinematics, input shaping and command smoothing. However, existed open-loop techniques cannot reject oscillations caused by external wind disturbances.

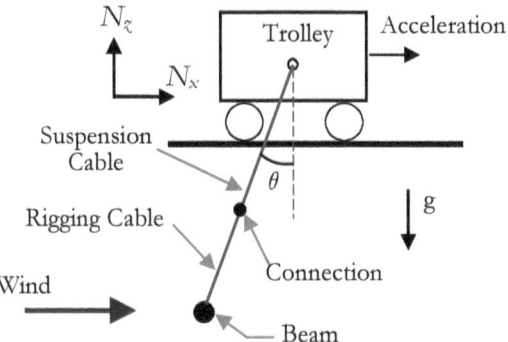

Figure 3.24 Model of a bridge crane with distributed-mass beams and wind gusts.

Another method has combined input shaping with feedback control to suppress oscillations caused by both human-operator commands and external disturbances [56]. The payload swing caused by human-operator commands was reduced by input shaping techniques, and that induced by wind disturbances was damped by a low-authority feedback controller.

Accurately sensing payload oscillations is a barrier for the feedback controller, and rejecting the external disturbance is an obstacle toward the application of the previous presented open-loop control scheme. This section will report a new method to control payload oscillations caused by both human-operator commands and wind disturbances without measuring the payload states on-the-fly.

3.4.1 Model

A schematic representation of a bridge crane with distributed-mass beams and wind gusts is shown in Figure 3.24. The trolley moves along the bridge in the x direction. A uniformly distributed-mass beam of mass, m_p, and length, l_p, hangs below the trolley by a massless suspension cable of length, l_b, and two massless rigging cables of length, l_r. A wind force, w_f, acts on the payload in the x direction. The wind travels perpendicular to the initial direction of the edge of the payload length and parallel to the bridge.

The payload twisting does not occur when the initial direction of the payload length is perpendicular to the wind direction and the bridge. The payload twisting dynamics is very complicated. Thus, the specified direction simplifies the dynamic model. The acceleration of the trolley, a, and the wind force, w_f, are inputs to the model. The swing angle of the suspension cable, θ, is the output. Motions of the payload are assumed to have no effects on the trolley motion due to the large mechanical impedance in the drive system. The damping ratio is approximately zero. The equation of the motion is derived using the Kane's method. The dynamic model relating the swing angle to the trolley acceleration, suspension cable length and the wind force is:

$$(l_h + \sqrt{l_v^2 - 0.25l_p^2}) \cdot \ddot{\theta} + 2\dot{l}_h\dot{\theta} + g\sin\theta = (a - \frac{w_f}{m_p}) \cdot \cos\theta, \tag{3.20}$$

Assuming a small angle approximation yields the linearized natural frequency:

$$\omega = \sqrt{g/(l_h + \sqrt{l_v^2 - 0.25l_p^2})}. \tag{3.21}$$

Thus, the natural frequency is dependent on the suspension cable length, rigging cable length and payload length. It is clear that the suspension cable length is fundamental, and the payload length also has some effects on the frequency.

Simulated payload oscillations resulting from both a 1.4 m driving motion and a 0.2 N wind disturbance are shown in Figure 3.25. The payload mass, suspension length, rigging length, and payload length were fixed at 0.3 kg, 1.0 m, 0.5 m, and 0.5 m, respectively. The wind gust caused payload oscillations at 0 s. Then the crane accelerated at 4.5 s, which induced another payload oscillations. When the crane decelerated later, additional oscillations were induced.

The system response includes three stages. The time frame before the trolley moves, in which wind gusts cause the payload swing, is defined as the first stage. The time frame when the trolley is in motion is defined as the transient stage. The time frame after the trolley is stopped is defined as the residual stage. The peak-to-peak deflection during the transient stage is referred as the transient deflection, while the peak-to-peak deflection during the residual stage is defined as the residual amplitude.

In Figure 3.25, oscillations caused by the wind are in phase with that caused by the acceleration. Therefore, oscillations become large during the transient stage. Additionally, oscillations caused by the deceleration and that during the transient stage are out of phase. Thus, oscillations during the residual stage decrease. The transient deflection and residual amplitude caused by both the wind and the original human-operator command were 31.5 cm and 16.4 cm, respectively. The simulated results indicate that the addition of wind disturbances makes the dynamics complicated and operating crane becomes challenging.

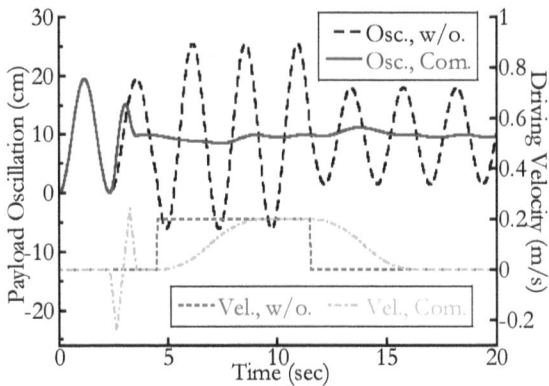

Figure 3.25 Oscillations resulting from driving motion and wind disturbance.

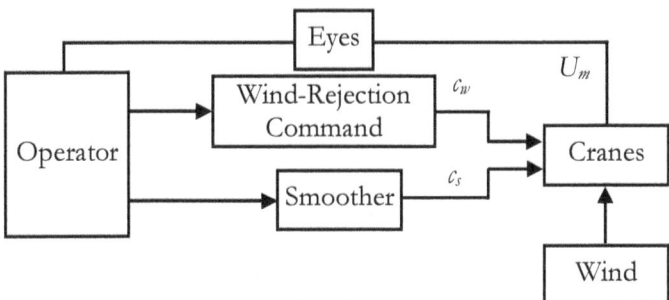

Figure 3.26 Combined control scheme.

3.4.2 Wind Rejection

A combined control technique is designed to control oscillations caused by wind disturbances and human-operator commands. Oscillations induced by wind disturbances are rejected by a wind-rejection command, and the payload swing generated by human-operator commands is eliminated by a smoother in Chapter 2. The combined control scheme is shown in Figure 3.26. The operator produces an original command, c_r, via the control interface. Then the original command (trapezoidal-velocity profile) filters through the smoother to create a smoothed command, c_s. A wind-rejection command, c_w, and the smoothed command, c_s, work together to move the trolley toward the desired position without inducing payload oscillations. The payload equilibrium angle, U_m, was utilized to design the wind-rejection command. Moreover, the equilibrium angle can be easily estimated by the operator with his/her eyes.

Oscillations around the equilibrium angle are assumed to be small for the crane dynamics shown in equation (3.20). During the crane motion, the suspension length does not change. Then the equation (3.20) can be simplified as:

$$\ddot{\theta} + \omega^2 \theta = \omega^2 (\frac{a}{g} - \frac{w_f}{m_p g}).$$

(3.22)

Therefore, equation (3.22) can be considered as a simplified model for a bridge crane with distributed-mass beams and wind gusts. The new equilibrium angle for this simplified model is given by:

$$U = \frac{w_f}{m_p g}.$$

(3.23)

The design of the wind-rejection command is very challenging for the crane dynamics shown in equation (3.20). Therefore, a disturbance-rejection command was obtained for a second-order system subject to external step disturbance. Then, the disturbance-rejection command was refined as a wind-rejection command for the bridge crane with distributed-mass beams and wind gusts. The resulting wind-rejection command will be used to test its performance on a crane dynamics shown in equation (3.20).

The harmonic response of the second-order system from the disturbance-rejection command, c, and the external step disturbance, D, is given by:

$$f(t) = \int_{\tau=0}^{+\infty} D \frac{\omega}{\sqrt{1-\zeta^2}} e^{-\zeta\omega(t-\tau)} \sin(\omega\sqrt{1-\zeta^2}(t-\tau))d\tau$$
$$+ \int_{\tau=0}^{+\infty} c(\tau) \frac{\omega}{\sqrt{1-\zeta^2}} e^{-\zeta\omega(t-\tau)} \sin(\omega\sqrt{1-\zeta^2}(t-\tau))d\tau$$

(3.24)

where ω is undamped natural frequency of the second-order system, and ζ is the corresponding damping ratio. The vibration amplitude of response (3.24) is:

$$A(t) = \frac{\omega}{\sqrt{1-\zeta^2}} e^{-\zeta\omega t} \sqrt{[S(\zeta,\omega,D)]^2 + [C(\zeta,\omega,D)]^2},$$

(3.25)

where

$$S(\zeta,\omega,D) = \int_{\tau=0}^{+\infty} c(\tau) e^{\zeta\omega\tau} \cos(\omega\sqrt{1-\zeta^2}\tau)d\tau - \zeta D / \omega,$$

(3.26)

$$C(\zeta,\omega,D) = \int_{\tau=0}^{+\infty} c(\tau) e^{\zeta\omega\tau} \sin(\omega\sqrt{1-\zeta^2}\tau)d\tau + \sqrt{1-\zeta^2} D / \omega.$$

(3.27)

48

The disturbance-rejection command, c, could cancel oscillations resulting from the external disturbance when equations (3.26) and (3.27) are limited to zero. A constraint must be satisfied to ensure that the velocity of the system can be limited to zero after the disturbance-rejection command:

$$\int_{\tau=0}^{\infty} c(\tau)d\tau = 0.\qquad(3.28)$$

Limiting equations (3.26) and (3.27) to zero and resulting from constraint (3.28) yield a disturbance-rejection command [59]:

$$c(\tau) = \begin{cases} A_1 e^{-\zeta_m \omega_m \tau} & , & \Delta \le \tau < (\Delta + T_m/8) \\ -A_2 e^{-\zeta_m \omega_m \tau} & , & (\Delta + T_m/8) \le \tau < (\Delta + T_m/4) \\ -A_1 e^{-\zeta_m \omega_m (\tau - T_m/4)} & , & (\Delta + T_m/4) \le \tau < (\Delta + 3T_m/8) \\ A_2 e^{-\zeta_m \omega_m (\tau - T_m/4)} & , & (\Delta + 3T_m/8) \le \tau \le (\Delta + T_m/2) \end{cases}\qquad(3.29)$$

where

$$A_1 = \frac{D_m \sqrt{1-\zeta_m^2}\{[\zeta_m + (\sqrt{2}-1)\sqrt{1-\zeta_m^2}] - [(\sqrt{2}-1)\zeta_m - \sqrt{1-\zeta_m^2}]e^{\zeta_m \pi/\sqrt{1-\zeta_m^2}}\}}{(2-\sqrt{2})(e^{\zeta_m \pi/\sqrt{1-\zeta_m^2}} + 1)}$$

, (3.30)

$$A_2 = \frac{D_m \sqrt{1-\zeta_m^2}\{[(\sqrt{2}-1)\zeta_m + \sqrt{1-\zeta_m^2}] - [\zeta_m - (\sqrt{2}-1)\sqrt{1-\zeta_m^2}]e^{\zeta_m \pi/\sqrt{1-\zeta_m^2}}\}}{(2-\sqrt{2})(e^{\zeta_m \pi/\sqrt{1-\zeta_m^2}} + 1)}$$

, (3.31)

$$\Delta = nT_m \qquad n = 0,1,2\cdots,\qquad(3.32)$$

D_m is the modeled external disturbance, and Δ is the initial time of the disturbance-rejection command. When the modeled damping ratio, ζ_m, modeled natural frequency, ω_m, modeled damped oscillation period, T_m, modeled external disturbance and the initial time of the disturbance-rejection command are correct, oscillations induced by the external disturbance will be limited to zero using the disturbance-rejection command, c. The duration of the disturbance-rejection command is one half of the damped oscillation period.

Disturbance-rejection command (3.29) and simplified dynamics (3.22-3.23) yield a wind-rejection command [59]:

$$c_w(\tau) = \begin{cases} -(\sqrt{2}+1)gU_m/2 & , & \Delta \le \tau < (\Delta + T_m/8) \\ (\sqrt{2}+1)gU_m/2 & , & (\Delta + T_m/8) \le \tau < (\Delta + 3T_m/8) , \\ -(\sqrt{2}+1)gU_m/2 & , & (\Delta + 3T_m/8) \le \tau < (\Delta + T_m/2) \end{cases}\qquad(3.33)$$

where U_m is the equilibrium angle for the bridge crane with distributed-mass beams and wind gusts.

The wind-rejection command is dependent on the modeled natural frequency, ω_m, equilibrium angle, U_m, and the initial time, Δ. However, a great wind force might cause actuator saturation. The wind force is assumed to be small such that the magnitude of the wind-rejection command is within the bound of the actuator. When operators have the visibility of the payload, both the equilibrium angle, U_m, and the initial time, Δ, can be estimated by eyes. The maximum deflection of the payload can be easily estimated by the human-operator with his/her eyes. Then half of the maximum deflection is corresponding to the equilibrium position of the payload. Meanwhile, the time for zero payload deflection is corresponding to the initial time, Δ. Therefore, both the initial time, Δ, and the equilibrium angle, U_m, do not need to be estimated on-the-fly.

Oscillations resulting from both wind disturbances and human-operator commands could be eliminated under the action of the wind-rejection command, c_n, and the smoothed command, c_s. Figure 3.25 also shows the effectiveness of the combined control scheme. The combined command drives the trolley with inducing minimal vibrations. With the combined command, the transient deflection and residual amplitude were suppressed to 15.0 cm and 0.4 cm, respectively. The simulated process clearly verifies the effectiveness of the combined control scheme.

3.4.3 Experimental Results

Two sets of experiments were performed to verify the effectiveness of the presented method on suppressing payload oscillations for variations of system parameters. The first set of experiments investigated the effect of variation in the wind forces. The payload length and mass were 0.4 m and 0.182 kg, respectively. Adjusting the distance between the fan and payload varies the wind force on the payload. As the distance increased, the equilibrium position of the payload and wind force decreased. The distances were set to 0.5, 0.6, 0.7, 0.8 and 0.9 m. The estimates of the new equilibrium position corresponding to the distances were approximately 10.23 cm, 9.80 cm, 9.33 cm, 8.80 cm and 8.16 cm, respectively. According to the equilibrium position equation, corresponding wind forces were 0.145 N, 0.139 N, 0.132 N, 0.125 N, and 0.116 N, respectively. The driving distance for the desired task was fixed at 0.8 m. The combined controller was designed when the modeled wind force was set to 0.132 N.

Experimental transient deflection and residual amplitude from those tests are shown in Figure 3.27. Without the controller, increasing wind force increased the transient deflection and residual amplitude. Experimental

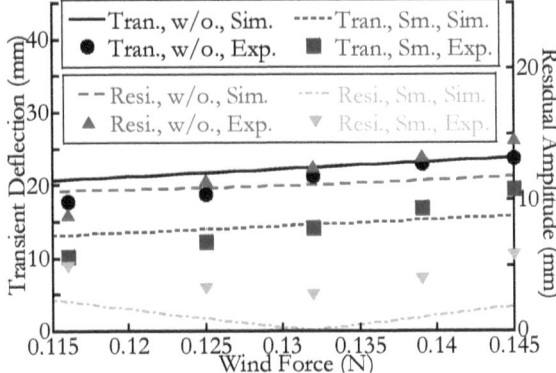

Figure 3.27 Experimental amplitudes induced by wind force.

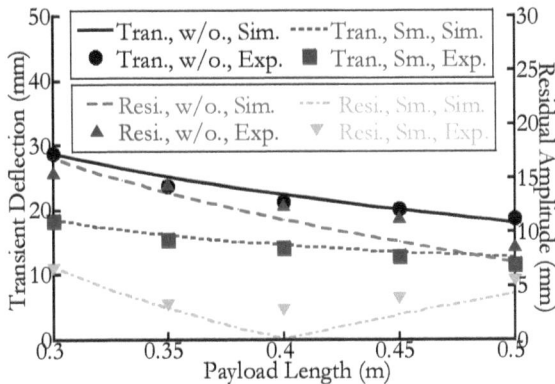

Figure 3.28 Experimental amplitudes induced by payload length.

results match the general trend as the simulated curve. With the combined controller, reduction in the transient deflection and residual amplitude was remarkable. The experimental residual amplitude was not eliminated to zero at the modeled wind force. This is because the actual system had estimation error of the initial time, while the initial time of the wind-rejection command in the simulation was correct. Increasing modeling errors decreased the performance of vibration reduction. The combined controller reduced the transient deflection by an average of 29.9 %, and residual amplitude by an average of 66.3 %. Therefore, the combined controller was effective at each of tested wind forces.

Another set of experiments was conducted to reduce payload swing from varying payload lengths as shown in Figure 3.28. Without the controller, the experimental transient deflection and residual amplitude decreased with increasing payload length. The combined control arithmetic was applied when the modeled payload length was set to 0.4 m. Both the transient deflection and residual amplitude had a dramatical reduction. The residual amplitude at the modeled payload length was not attenuated to zero because of estimation errors of the initial time of the wind-rejection command. The combined controller reduced the transient deflection by an average of 35.4 %, and the residual amplitude by an average of 65.2 %. Those experiments shown in Figures 3.27-3.28 clearly verified that the combined control technique can effectively suppress oscillations induced by both wind disturbances and human-operator commands.

3.5 Chapter Summary

This chapter described control of bridge cranes having single-pendulum dynamics, double-pendulum dynamics, and distributed-mass payload dynamics by using the three types of command smoother. The two-, three-, and four-pieces smoothers are effective to reduce payload oscillations for various working conditions and system parameters. Meanwhile, smoothers provide enough robustness to variations in the frequency. Simulations and experiments were used to verify the key dynamic behavior and the effectiveness of smoothers.

Chapter 4. Application in Tower Cranes

Tower cranes are broadly applied in construction throughout the world. Figure 4.1 shows a small-scale tower crane carrying a beam. Tower cranes transport payloads by translating a trolley in and out along the jib, slewing a jib about the tower, and hoisting a suspension cable. Nonetheless, payload oscillations induced by motions commanded by the human operator are a major drawback for efficient and safe manipulation. Operators usually address oscillatory problems manually by driving slowly and waiting for undesirable oscillations to decay.

Payloads hang below the hook by rigging cables, which are utilized for transporting bulky objects. The dynamics are very complex due to the payload swing and twisting about the rigging cables. Additionally, operation tasks may be more challenging because even skilled operators cannot control the payload twisting manually. Therefore, there is an essential to research twisting dynamics and control in tower cranes moving bulky payloads.

Many scientists have studied on the dynamics and control of tower cranes with point-mass payloads. The payload twisting cannot be captured in the point-mass payload model because of neglecting the payload size. The control methods for tower cranes with point-mass payloads include closed-loop controllers and open-loop controllers. Closed-loop controllers measure or estimate the payload swing to reduce oscillations in a feedback loop, including gain-scheduled control [60-61], fuzzy control [62-63], H∞ control [64], neural network control [65-66], sliding mode control [67-68], LQR control [69], path-following control [70], adaptive control [71-72], and predictive control [73]. However, accurately measuring the payload swing on-the-fly is difficult [74]. Open-loop controllers modify the prescribed commands for swing suppression, including inverse kinematics [75], smooth commands [76], and input shaping [5, 35, 38, 52]. Nevertheless, all the above-mentioned works have focused on tower cranes with point-mass payload dynamics.

The payload twisting dynamics in bridge cranes have been studied in Chapter 3. Of course, the dynamics of the payload twisting of bridge cranes is similar to that of tower cranes during pure radial motions. However, the dynamics of the payload twisting in tower cranes during slewing motions is much more complicated. This Chapter presents a model of tower cranes transporting distributed-mass beams. The dynamic behavior of the payload twisting in tower cranes during slewing motions is also studied. A demonstration that the smoother described above is also effective on oscillation reduction for these types of cranes.

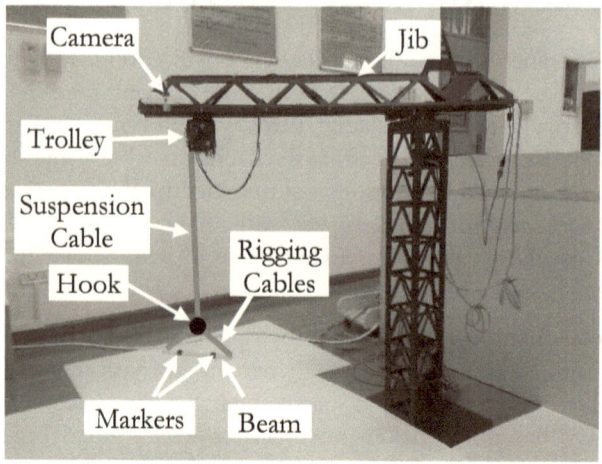

Figure 4.1 A testbed of a tower crane transporting a distributed-mass beam.

4.1 Modeling

Figure 4.2 shows a schematic representation of a tower crane transporting a distributed-mass beam. While a trolley moves radially with a position of r along the jib, a jib arm rotates by an angle of θ about the tower. A hook of mass, m_h, is attached to the trolley by a massless suspension cable of length, l_s. Two massless rigging cables of length, l_r, hangs below the hook and supports a uniformly distributed-mass beam of mass, m_p, and length, l_p. The payload twist angle, γ, is defined as the angle between the radial jib direction and the long axis of the beam.

The model assumes that the motion of the tower crane is unaffected by motion of the hook and payload because of the high-ratio geared drives and small swing. It is also assumed that the damping ratio is approximately zero, the hook is modeled as a point mass, and the length of the suspension cable does not change during the motions. The angular acceleration of the jib and acceleration of the trolley are inputs to the model. The swing angles of the suspension cable a_x and a_y, the swing angles relative to the suspension cable β_x and β_y, and the payload twist angle γ, are the outputs.

The nonlinear equations of motion for the model in Figure 4.2 were derived by using Kane's method. The equations for the swing angles and the payload twist angle are:

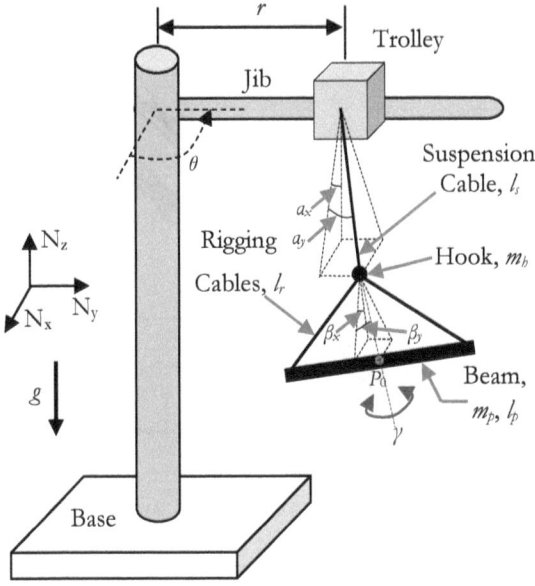

Figure 4.2 Model of a tower crane carrying a distributed-mass beam.

$$M \cdot \begin{pmatrix} \ddot{\alpha}_x \\ \ddot{\alpha}_y \\ \ddot{\beta}_x \\ \ddot{\beta}_y \\ \ddot{\gamma} \end{pmatrix} + f\begin{pmatrix} \alpha_x, \alpha_y, \beta_x, \beta_y, \gamma, \theta, r, \\ \dot{\alpha}_x, \dot{\alpha}_y, \dot{\beta}_x, \dot{\beta}_y, \dot{\gamma}, \dot{\theta}, \dot{r}, \\ \ddot{\theta}, \ddot{r} \end{pmatrix} = 0, \tag{4.1}$$

where the mass matrix is M, and the column matrix of gravity terms, centrifugal and Coriolis terms, and control input terms is f.

A trapezoidal-velocity profile was used to drive both the model (4.1) and the small-scale tower crane shown in Figure 4.1 through a slew of 80 degrees in order to verify the nonlinear equations of motion. The jib accelerates at time zero. Then, the jib decelerates 4 seconds later. Both the acceleration and deceleration induce both the payload swing and twisting. The maximum slewing speed between the acceleration and deceleration is $20°/\text{s}$. Figure 4.3 shows the experimental and simulated responses of the payload swing. The position deflection of the mass center of the payload

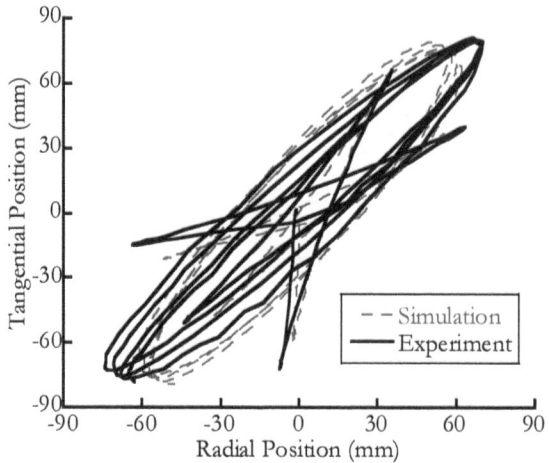

Figure 4.3 Experimental verification of payload swing.

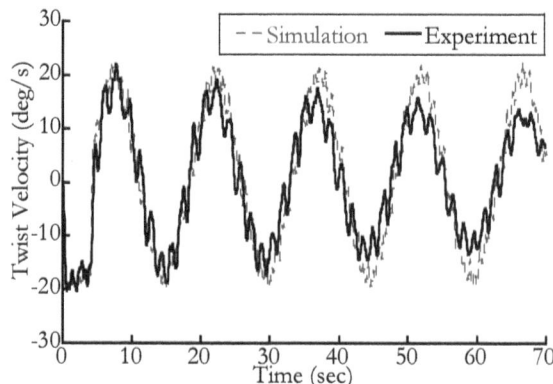

Figure 4.4 Experimental verification of payload twisting.

relative to the trolley is defined as the payload swing displacement. In addition, Figure 4.4 shows results of the angular velocity of the payload twist. The experimental data shows fairly good agreement with the nonlinear simulations. The difference between the simulated and experimental results is caused by the friction and air drag. The decay of the experimental response revealed such difference, which is most easily seen in the twist result in Figure 4.4.

4.2 Dynamics

The dynamics of tower cranes during pure radial motions are similar to that of bridge cranes. The payload twisting cannot be excited by pure radial motions in tower cranes when the initial twist angle is limited to zero. Thus, the simplified equations during the pure radial motions are:

$$
\begin{bmatrix} [cR^2 + l_s^2 + c(l_y^2 + l_s^2 + 2l_yl_s\cos(\beta_y))] & c[R^2 + l_y(l_y + l_s\cos(\beta_y))] \\ [R^2 + l_y(l_y + l_s\cos(\beta_y))] & [R^2 + l_y^2] \end{bmatrix} \cdot \begin{pmatrix} \ddot{\alpha}_y \\ \ddot{\beta}_y \end{pmatrix}
$$

$$
+ \begin{bmatrix} -cl_yl_s\sin(\beta_y) \\ 0 \end{bmatrix} \cdot \dot{\beta}_y^2 + \begin{bmatrix} 0 \\ l_yl_s\sin(\beta_y) \end{bmatrix} \cdot \dot{\alpha}_y^2 + \begin{bmatrix} -2cl_yl_s\sin(\beta_y) \\ 0 \end{bmatrix} \cdot \dot{\alpha}_y\dot{\beta}_y
$$

$$
+ \begin{bmatrix} l_s(1+c)\sin(\alpha_y) + cl_y\sin(\alpha_y + \beta_y) \\ l_y\sin(\alpha_y + \beta_y) \end{bmatrix} \cdot g
$$

$$
+ \begin{bmatrix} [l_s\cos(\alpha_y) + cl_s\cos(\alpha_y) + cl_y\cos(\alpha_y + \beta_y)] \\ [l_y\cos(\alpha_y + \beta_y)] \end{bmatrix} \cdot \ddot{r} = 0
$$

, 　　(4.2)

where the gravitational constant is g, and the ratio of the payload mass to the hook mass is c. The coefficients R and l_y are given by:

$$
R = l_p / (2\sqrt{3}), \tag{4.3}
$$

$$
l_y = \sqrt{l_r^2 - 0.25l_p^2}. \tag{4.4}
$$

A linearized model can be derived from the abovementioned simplified model to approximate the pure radial motions by assuming small oscillations around the equilibrium position. Then the linearized natural frequencies of the payload swing during pure radial motions are given by:

$$
\omega_{2,1}^2 = \frac{g(c+1)}{2l_s}(u \pm v), \tag{4.5}
$$

where

$$
u = \frac{l_y^2 + l_sl_y + R^2}{l_y^2 + (c+1)R^2}, \tag{4.6}
$$

$$
v = \sqrt{u^2 - \frac{4l_sl_y}{(c+1)(l_y^2 + (c+1)R^2)}}. \tag{4.7}
$$

The natural frequencies of the payload swing during slewing motions can also be estimated by equation (4.5). This is because the payload swing exhibits weakly nonlinear dynamic behavior, especially when tower cranes move slowly. The swing frequencies depend on the mass ratio, suspension cable length, rigging cable length, and payload length. The mass ratio, cable length, and payload size have larger effects on the second-mode swing frequency than the first-mode swing frequency. The payload size and mass ratio only have minor impacts on the first-mode swing frequency. Meanwhile, increasing cable length decreases the first-mode swing frequency slightly, and decreases the second-mode swing frequency sharply.

The hook mass is set to be zero, and the swing angles are assumed to be small around the equilibrium point. Therefore, a simplified model for the twisting dynamics can be derived:

$$\ddot{\gamma} + \left[\alpha_y \sin(2\gamma) + \alpha_x - \alpha_x \cos(2\gamma) \right] \dot{\alpha}_x \dot{\theta} + \left[\alpha_x \sin(2\gamma) - 2\alpha_y \sin^2 \gamma \right] \dot{\alpha}_y \dot{\theta}$$

$$+ \left[\alpha_x \sin \gamma + \alpha_y \cos \gamma \right] \left[\alpha_y \sin \gamma - \alpha_x \cos \gamma \right] \cdot (\dot{\theta})^2 + \ddot{\theta} - \alpha_y \cdot \ddot{\alpha}_x = 0$$

$$(4.8)$$

The twist acceleration is dependent on slewing motions and payload swing. The slewing motions is external excitation, while the payload swing is parametric excitation. The twisting response is similar to a harmonic motion in the residual stage (no external excitation). The payload twists about the swing direction back and forth. This is because the payload swing results in the twist acceleration. The magnitudes of the twist acceleration are dependent on the amplitude of the payload swing. The sign of the twist acceleration depends on the payload position. Zero swing of the payload causes zero acceleration of the payload twisting. The inertia effect will rotate the payload in one direction under conditions of constant twist angular velocity.

The twist frequency for various amplitudes and first-mode frequency of the payload swing is shown in Figure 4.5, in which the amplitude and frequency of the swinging changes independently. The first-mode swing frequency is larger than the twist frequency. Both swing amplitude and frequency have a great impact on the twist frequency. The twist frequency increases as swing amplitude and frequency increase. This effect can be physically interpreted as the interference between the sign and magnitude of the twist acceleration. As the amplitude and frequency of the payload swing increase, the payload rotates faster before changing the sign of the twist acceleration. Furthermore, the twist frequency is limited to zero in the case of zero swing because zero payload swing results in zero twist acceleration. Thus, the payload will spin in one direction on condition of a

58

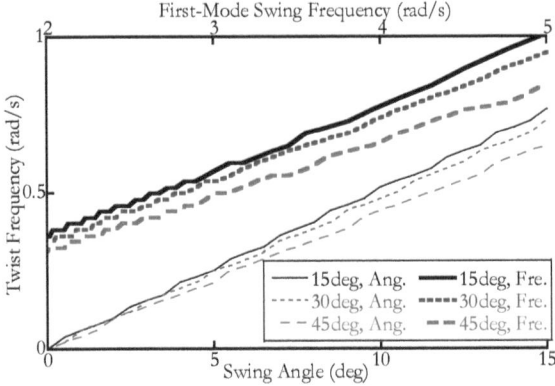

Figure 4.5 Twist frequency.

constant angular velocity of the payload twist. Therefore, decreasing twist frequency results from reduction of both the amplitude and frequency of the swing.

The initial twist angle also has a great effect on the payload twisting dynamics. The payload twisting dynamics for initial twist angles of 15 deg, 30 deg, and 45 deg are also shown in Figure 4.5. The twist frequency decreases with increasing the initial twist angle. This effect can also be interpreted as the interference between the sign and magnitude of the twist acceleration. When the initial twist angle increases, the payload rotates further before changing the sign of the twist acceleration. Because the complicated dynamical behavior of the payload twisting is sensitive to the initial twist angle, the payload twisting displays strongly nonlinear dynamical behavior.

4.3 Experimental Results

The testbed shown in Figure 4.1 was used to verify the dynamical behavior of the model and effectiveness of the smoother. The position of the trolley along the jib was set to be 75 cm. A tennis ball and a slender beam served as the hook and payload, respectively. Both the suspension cable and rigging cables were made of Dyneema fishing line. The hook mass, payload mass, payload length, suspension length, and rigging length were fixed at 32 g, 155 g, 29.7 cm, 54 cm and 16 cm, respectively.

To record the displacements of two markers on the payload, the jib mounted to a camera, which captured video at 30 frames per second. The displacements of two markers were used to measure the payload swing displacement and twist angle. Averaging the displacements of the two

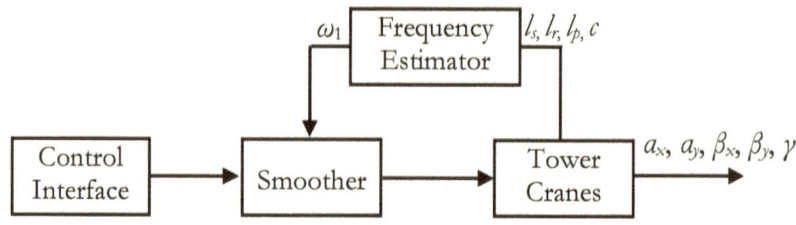

Figure 4.6 Control architecture for tower cranes.

markers derived the payload swing displacement. The twist angular velocity can be estimated by using the inverse tangent function.

The experimental control architecture is shown in Figure 4.6. A baseline trapezoidal-velocity profile is produced via the control interface. The command is then modified by the two-pieces smoother to produce a smoothed command for slewing the jib. The design frequency for the smoother was estimated by the suspension cable length, l_s, rigging cable length, l_r, payload length, l_p, and mass ratio, c. The design damping ratio is fixed at zero in the experiments.

The experimental response of the payload swing caused by the baseline trapezoid and smoothed commands are shown in Figure 4.7. The jib was slewed 80 degrees using a maximum velocity of 20 deg/s. The residual amplitude is defined as the maximum deflection after the jib stops. The residual amplitude of the uncontrolled swing was 106 mm, while residual amplitude with the smoother was only 9.9 mm.

Figure 4.8 shows the slewing velocity profile and the experimental angular velocity of the payload twist. The smooth transitions between boundary conditions for the smoothed velocity profile reduce swing and twisting of the payload. Without the controller, the residual amplitude of the twist angular velocity was 21.9 deg/s. The residual amplitude with the smoother was 6.8 deg/s. Meanwhile, the experimental frequency of large-amplitude low-frequency twist oscillations without the controller was 0.43 rad/s, while that with the smoother was only 0.1 rad/s. Thus, the smoother benefits reduction of the twist frequency. Such a reduction would largely improve the safety of a real crane transporting a heavy and bulky payload.

Figure 4.9 shows the experimental and simulated results of the residual amplitudes of the payload swing. Peaks and troughs arose in the amplitude of the payload swing when using the unsmoothed trapezoidal commands. The design frequency and damping ratio of the smoother were set to be 4.032 rad/s and 0, respectively. The smoother eliminated most of the

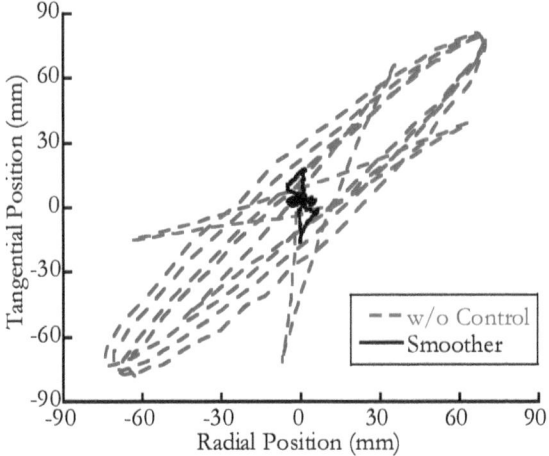

Figure 4.7 Experimental responses of payload swing.

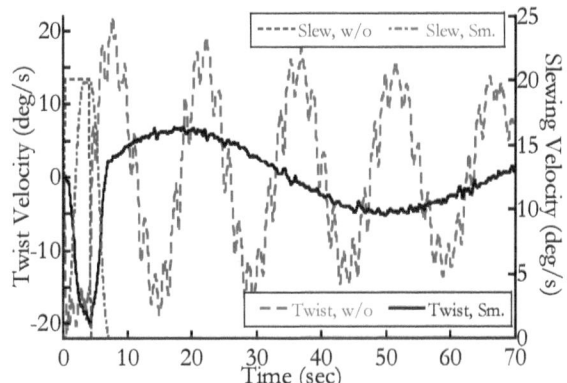

Figure 4.8 Experimental responses of payload twisting.

payload swing such that the controlled results were nearly independent of the slewing motions. The experimental results with the smoother were somewhat worse than the simulated results. This is because small modeling errors and uncertainty exist in the overall system.

The results of the amplitude and frequency of the payload twisting are shown in Figure 4.10. By using the fast Fourier transform, the twist frequency was derived. As the rotating distance changed, peaks and troughs also arose in both the amplitude and frequency of the payload twisting. The magnitude of the twist acceleration reached a large value when the swing amplitude reached a maximum. Therefore, peaks in the twist amplitude and frequency occurred. As the swing amplitudes increase, the amplitude and frequency of the payload twisting increase. With the smoother, the

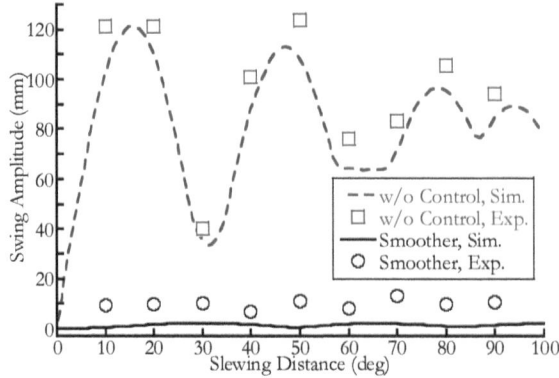

Figure 4.9 Experimental responses of payload swing as a function of slewing distances.

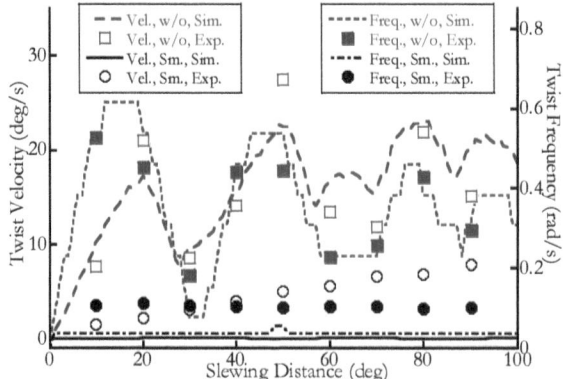

Figure 4.10 Experimental responses of payload twisting as a function of slewing distances.

experimental twist velocity increases as the slewing distance increases, because the model is undamped while the real system has some small amount of damping. The small damping corrupts the oscillation reduction for long slewing distances. However, the experimental data follow the general shape of the simulated results. The smoother attenuated the experimental swing amplitude, twist amplitude, and twist frequency by an average of 89.8%, 71.7%, and 71.0%, respectively. The experimental data verified the theoretical dynamical behavior and the effectiveness of the two-pieces smoother.

Figure 4.11 and Figure 4.12 show the experimental payload swing and twisting resulting from different modeling errors in the frequency. The ratio of the modeled frequency to the real frequency is defined as the normalized frequency. The design frequency of 4.032 rad/s corresponds to the

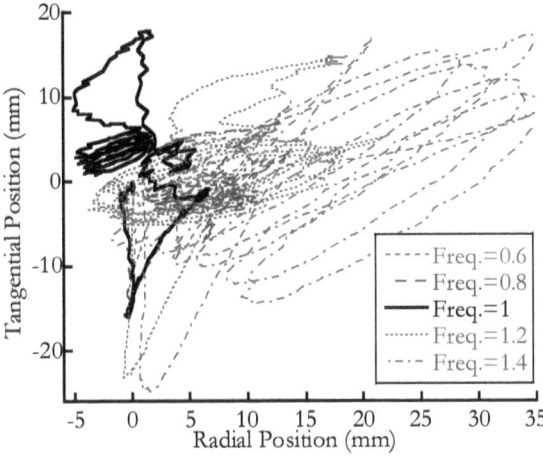

Figure 4.11 Experimental responses of payload swing for various modeling errors in the frequency.

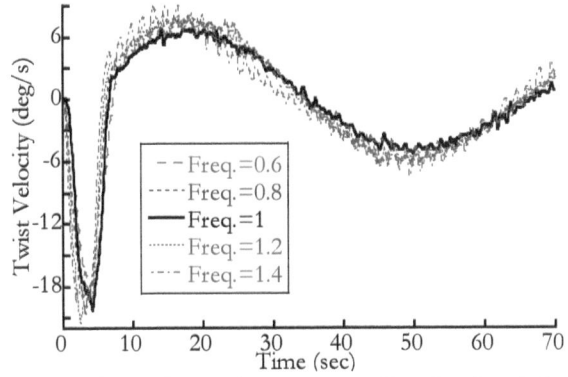

Figure 4.12 Experimental responses of payload twisting for various modeling errors in the frequency

normalized frequency of one. In the case of normalized frequency of one, the residual amplitudes of the swing and twisting were 9.9 mm and 6.8 deg/s, respectively. The residual amplitudes of the swing and twisting in the case of the normalized frequency of 0.8 (corresponding to the design frequency of 3.226 rad/s) were 11.1 mm and 8.0 deg/s, respectively. The residual amplitudes of the swing and twisting for the normalized frequency of 1.2 (corresponding to the design frequency of 4.839 rad/s) were 18.6 mm and 8.3 deg/s, respectively. The residual amplitudes of the swing and twisting for the frequency of 0.6 were 10.1 mm and 7.6 deg/s, and those for the normalized frequency of 1.4 were 36.8 mm and 9.0 deg/s.

The experimental results clearly indicate that the payload swing increases with increasing modeling error. The payload swing exhibits insensitivity to the modeling error at low normalized frequencies due to the low-pass filtering effect. The experimental results show that the controlled twisting response is insensitive to modeling errors in the frequency. The smoother suppresses both the payload swing and twisting to a very low level, and provides good insensitivity to modeling errors in the frequency.

4.4 Chapter Summary

This chapter described a dynamic model of tower cranes transporting distributed-mass payloads. The complicated nonlinear dynamics were investigated for a wide range of slewing motions. The smoother was successful in suppressing both the payload swing and twisting. The dynamical behavior of the nonlinear model and the effectiveness of the two-pieces smoother were verified from a testbed of a small-scale tower crane transporting a distributed-mass beam.

Chapter 5. Application in Flexible Manipulators

The study of flexible link manipulators has been actuated by demands of light-weight robotic systems and space applications. However, flexible link manipulators suffer from undesired oscillations caused by operator-commanded motions. The detrimental effects degrade positioning accuracy, capable operating speeds, and safety. Therefore, it is essential to control unwanted oscillations in flexible link manipulators.

5.1 Linear Dynamics

Broad attention have been attracted on modeling and dynamics of flexible link manipulators [77-84]. Dynamical analysis indicates that single-link flexible manipulator includes an infinite number of vibration modes. The first one is the fundamental mode, but high modes might have some effects. Therefore, designing a control system for reducing vibrations of all modes is essential.

Hundreds of papers reported vibration control for flexible link manipulators. The control schemes can be broken into two categories: feedback control and open-loop control. The feedback control strategies use measurements of the flexible link to control vibrations in a closed loop. The feedback control methods include the proportional-integral-derivative control [85-86], delayed feedback control [87], positive position feedback control [88], linear quadratic regulator [89], adaptive control [90-91], sliding-mode control [92-93], fuzzy control [94-95], and artificial neural networks based control [96-97]. The open-loop control strategies filter inputs to produce a desirable motion that results in minimal vibrations. The open-loop control methods include the optimal trajectory planning [98-100], and input shaping [101-104]. More literatures for dynamics and control of flexible link manipulators might be found from the review [105-109].

Many works have been directed at controlling multi-mode vibrations for single-link flexible manipulators. However, measuring high-mode vibrations is an obstacle towards the feedback-control application. Meanwhile, vibrations of high modes, whose frequencies lie outside of the bandwidth of the actuator or sensor, are challenging to be controlled by the feedback controller. The input shaper is designed for each modes of the flexible link independently, and then convolving them together. However, high-mode frequencies are challenging to measure or estimate.

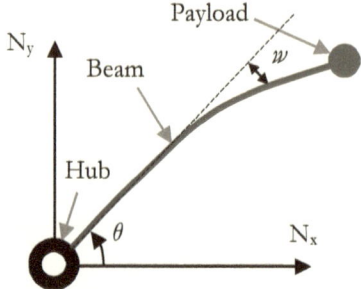

Figure 5.1 Model of a single-link flexible manipulator.

Additionally, convolving an infinite number of input shapers is impossible. Thus, previously presented control methods are focused on reducing vibrations for the first few modes.

5.1.1 Modeling

A schematic representation of a single-link flexible manipulator is shown in Figure 5.1. The hub inputs an angular displacement, θ. A flexible beam of length, l_b, is attached to the hub, and supports a payload of mass, m_p. Ideally, responses of the payload and beam track the angular displacement, θ. However, such flexible structure often causes unwanted vibrations of the payload and beam.

The hub acceleration is the input to the model. The output is the deflection, $w(x, t)$, of a point along the beam at a distance, x, from the hub. It is assumed that the motion of the beam and payload does not affect the motion of the hub because of the large mechanical impedance in the drive system. The force equation of the motion is given by:

$$\rho dx \cdot \frac{\partial^2 w(x,t)}{\partial t^2} + \rho dx \cdot x \ddot{\theta}(t) + dV(x, t) = 0 ,\qquad (5.1)$$

where ρ is the linear mass density of the beam, dx is an infinitely small change in the x, $w(x, t)$ is the deflection of a point along the beam at a distance, x, from the hub, $V(x, t)$ is the shearing force in the beam, and the differential $dV(x, t)$ represents an infinitely small change in the $V(x, t)$. The moment equation of the motion is given by:

$$dM(x, t) - V(x, t) \cdot dx - dV(x, t) \cdot dx - \rho dx \cdot x \cdot \ddot{\theta}(t) \cdot \frac{1}{2} dx = 0, \quad (5.2)$$

where $M(x, t)$ is the bending moment in the beam, and the differential $dM(x, t)$ represents an infinitely small change in the bending moment, $M(x, t)$.

From the theory of the beam bending, the relationship between bending moment and beam deflection satisfies:

$$M(x,t) = EI \frac{\partial^2 w(x,t)}{\partial x^2},$$ (5.3)

where I is the moment of inertia of the beam cross section, and E is Young's modulus. Substituting equations (5.2) and (5.3) into equation (5.1), and ignoring terms involving high powers in dx produces:

$$EI \frac{\partial^4 w(x,t)}{\partial x^4} + \rho \frac{\partial^2 w(x,t)}{\partial t^2} = -\rho \cdot x \cdot \ddot{\theta}(t),$$ (5.4)

The boundary conditions are given by:

$$w(x,t)|_{x=0} = 0,$$ (5.5)

$$\frac{\partial w(x,t)}{\partial x}|_{x=0} = 0,$$ (5.6)

$$\frac{\partial^2 w(x,t)}{\partial x^2}|_{x=l_b} = 0,$$ (5.7)

$$\frac{\partial}{\partial x}[EI \frac{\partial^2 w(x,t)}{\partial x^2}]|_{x=l_b} = m_p \frac{\partial^2 w(x,t)}{\partial t^2}|_{x=l_b}.$$ (5.8)

Using the mode superposition method, the deflection of the beam can be assumed as:

$$w(x,t) = \sum_{k=1}^{+\infty} \varphi_k(x) \cdot q_k(t),$$ (5.9)

where $\varphi_k(x)$ is the mode shape of the kth mode, and $q_k(t)$ is the time-dependent function of the kth vibration mode. The mode shape satisfies the following equations:

$$\varphi_k(x)|_{x=0} = 0,$$ (5.10)

$$\frac{\partial \varphi_k(x)}{\partial x}|_{x=0} = 0,$$ (5.11)

$$\frac{\partial^2 \varphi_k(x)}{\partial x^2}|_{x=l_b} = 0,$$ (5.12)

$$EI \frac{\partial^3 \varphi_k(x)}{\partial x^3}|_{x=l_b} + m_p \omega_k^2 \varphi_k(x)|_{x=l_b} = 0,$$ (5.13)

where ω_k is the natural frequency of the kth mode. Solving equations (5.10)-(5.13) yields the frequency ω_k, and the mode shape $\varphi_k(x)$:

$$\omega_k = (\beta_k l_b)^2 \sqrt{\frac{EI}{\rho l_b^4}}, \tag{5.14}$$

$$\cos(\beta_k l_b)\cosh(\beta_k l_b) + 1 = \\ h \cdot \beta_k l_b \cdot [\sin(\beta_k l_b)\cosh(\beta_k l_b) - \cos(\beta_k l_b)\sinh(\beta_k l_b)]' \tag{5.15}$$

$$\varphi_k(x) = C\phi_k(x), \tag{5.16}$$

$$\phi_k(x) = \sin(\beta_k x) - \sinh(\beta_k x) - r\cos(\beta_k x) + r\cosh(\beta_k x), \tag{5.17}$$

$$r = \frac{\sin(\beta_k l_b) + \sinh(\beta_k l_b)}{\cos(\beta_k l_b) + \cosh(\beta_k l_b)}, \tag{5.18}$$

where C is a constant, and h is the ratio of the payload mass to the beam mass. The closed-form solution for the coefficients, β_k and r, is challenging, but the numerical solution can be derived. Substituting equation (5.9) into equation (5.4) obtains:

$$EI\sum_{k=1}^{+\infty}[\frac{\partial^4\varphi_k(x)}{\partial x^4}q_k(t)] + \rho\sum_{k=1}^{+\infty}[\varphi_k(x)\frac{\partial^2 q_k(t)}{\partial t^2}] = -\rho x\ddot{\theta}(t). \tag{5.19}$$

Multiplying by $\varphi_k(x)$, and integrating over the length of the beam yield:

$$\frac{\partial^2 q_k(t)}{\partial t^2} + \omega_k^2 q_k(t) = -\frac{\gamma_k}{C\alpha_k}\ddot{\theta}(t), \tag{5.20}$$

where

$$\gamma_k = \int_0^{l_b} x\phi_k(x)dx, \tag{5.21}$$

$$\alpha_k = \int_0^{l_b} \phi_k(x)\phi_k(x)dx. \tag{5.22}$$

Equation (5.20) could be changed with the inclusion of proportional damping:

$$\frac{\partial^2 q_k(t)}{\partial t^2} + 2\zeta_k\omega_k\frac{\partial q_k(t)}{\partial t} + \omega_k^2 q_k(t) = -\frac{\gamma_k}{C\alpha_k}\ddot{\theta}(t), \tag{5.23}$$

where ζ_k is the damping ratio of the kth mode. The Laplace transform of the beam deflection at the distance, x, resulting from (5.23) is:

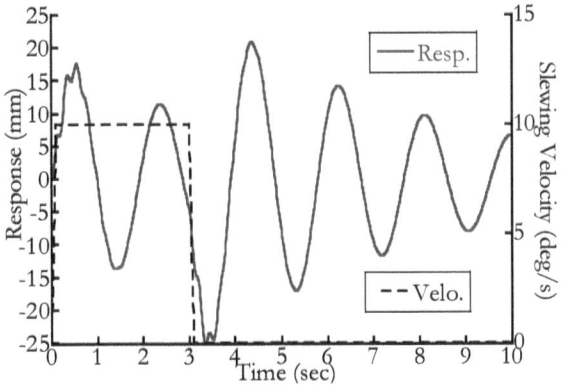

Figure 5.2 Simulated response for a driving distance of 30°.

$$w_x(s) = \sum_{k=1}^{+\infty} \frac{-\phi_k(x) \cdot \gamma_k}{\alpha_k \cdot (s^2 + 2\zeta_k \omega_k s + \omega_k^2)} \cdot \ddot{\theta}(s).$$

(5.24)

The model (5.24) includes an infinite number of vibrational modes. The deflection of the beam is a sum of response for each of an infinite number of vibration modes.

A trapezoidal velocity profile (bang-coast-bang acceleration) is applied to drive the hub. The Young's modulus, moment of inertia, linear mass density of the beam, damping ratio, maximum driving velocity, and maximum driving acceleration in the simulation were 2.06×10^5 MPa, 3.449 mm⁴, 0.3143 kg/m, 0.06, 10°/s, and 100°/s², respectively. When the hub moves, the maximum peak-to-peak deflection is referred to as the transient deflection. After the hub stops, the maximum peak-to-peak deflection is defined as the residual amplitude. A simulated response for a driving distance of 30° is shown in Figure 5.2 when the hub moves between 0s and 3s. The transient deflection is 31.2 mm, while the residual amplitude is 46.1 mm.

Residual amplitudes of the first mode and high modes for varying normalized distance, x/l_b, are shown in Figure 5.3. The beam length, driving distance, and mass ratio were fixed at 95 cm, 19°, and 0.5, respectively. The residual amplitude of the first mode increases as the normalized distance increases. As the normalized distance changes, a peak and a trough occur in the high mode, and the peak occurs near the middle point of the beam. The residual amplitude of the first mode at the middle point of the beam was 9.84 mm, while corresponding magnitude of high modes was 4.43 mm. When the measurement point locates near the middle point of beam, the high modes cause large vibrations.

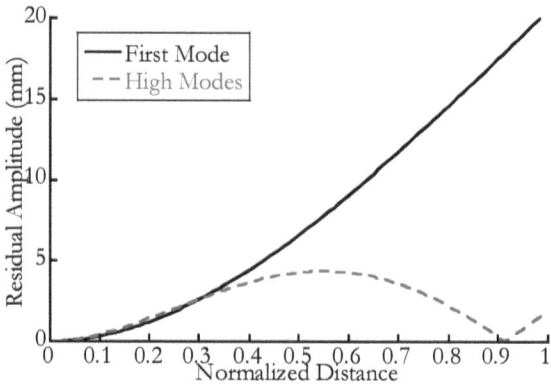

Figure 5.3. Residual amplitudes against normalized distance.

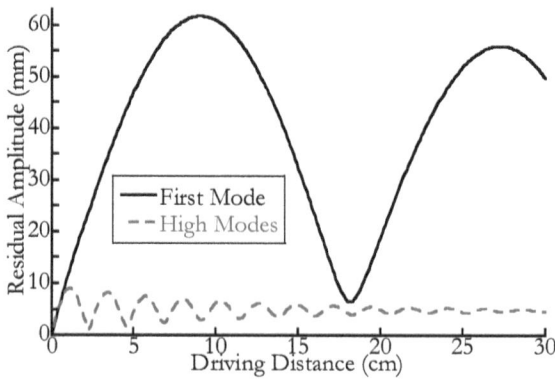

Figure 5.4 Residual amplitudes of the deflection against driving distance.

Residual amplitudes of the first mode and high modes for various driving distances are shown in Figure 5.4. The normalized distance, beam length, and mass ratio were fixed at 0.5, 95 cm, and 0.5, respectively. Peaks and troughs in the first mode and high modes exist as the driving distance changes. When vibrations caused by the acceleration and deceleration were in phase, peaks arose. When vibrations caused by the acceleration and deceleration were out of phase, troughs occurred. Peaks and troughs were spaced farther apart for the first mode. This is because the first mode had the low natural frequency. The average residual amplitudes of the first mode and high modes were 32.9 mm and 4.3 mm, respectively. The ratio of the average amplitude of high modes to that of the first mode was 13.1%. Therefore, the theoretical finding is that high modes have some effects in some specified cases.

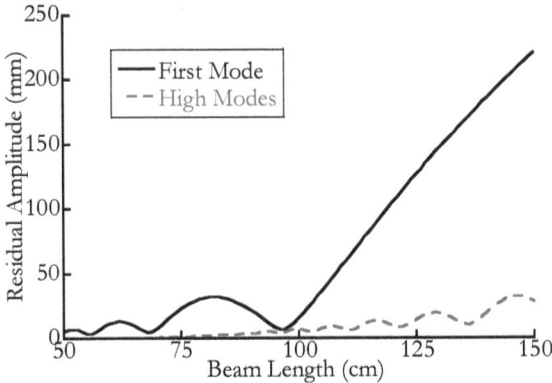

Figure 5.5 Residual amplitudes of the deflection against beam length.

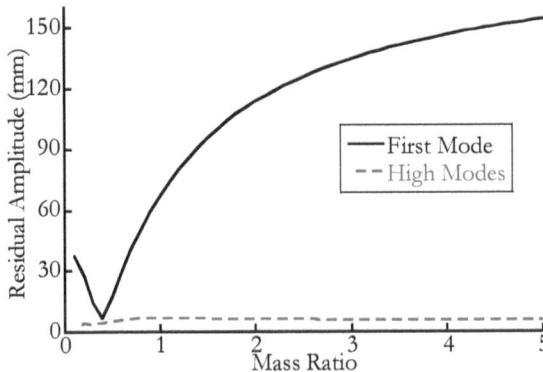

Figure 5.6 Residual amplitudes of the deflection against mass ratio.

Residual amplitudes of the first mode and high modes for varying beam lengths are shown in Figure 5.5. The normalized distance, driving distance, and mass ratio were fixed at 0.5, 19°, and 0.5, respectively. Peaks and troughs arose due to the interference between vibrations caused by the acceleration and deceleration. Residual vibrations of the first mode and high modes increase with increasing the beam length. The high modes cause large vibrations for long beams.

Residual amplitudes of the first mode and high modes for various mass ratios are shown in Figure 5.6. The normalized distance, driving distance, and beam length were fixed at 0.5, 19°, and 95 cm, respectively. Residual amplitudes of the first mode decreased as the mass ratio increased before 0.5. After this point, residual vibrations increased. Residual vibrations of high modes increase slightly as the mass ratio increases. Therefore, the mass ratio does not have large effects on the high-mode dynamics.

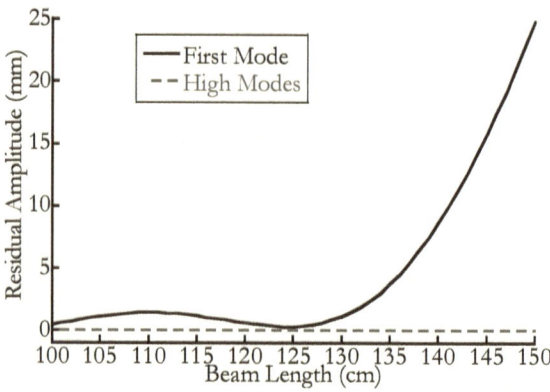

Figure 5.7 Simulated residual amplitudes against beam length.

Simulated result indicates that high modes have large effects when the middle point of the beam is chosen for measurement and the beam length is large. Therefore, it is essential to design a control system that can effectively reduce vibrations induced by total modes.

5.1.2 Numerical Verification

The four-pieces smoother presented in the Chapter 2 will be used to suppress infinite mode vibrations in flexible single-link manipulators. The four-pieces smoother exploits the first-mode frequency. However, the first-mode frequency may not be known or change over time in some specified cases. Thus, this section will study the numerical verification of the four-pieces smoother for modeling errors in the first-mode frequency.

Simulated residual amplitudes for varying beam lengths are shown in Figure 5.7. The normalized distance, x/l_b, driving distance, and mass ratio were fixed at 0.5, 19°, and 0.5, respectively. The four-pieces smoother was designed when the modeled length of the beam was 125 cm. The real beam length ranged from 100 cm to 150 cm. The smoother limited first-mode vibrations to near zero before the 125 cm because of the low-passing effect of the smoother. As the beam length increased from 125 cm, residual amplitudes with the smoother increased sharply. This is because the smoother has relatively narrow insensitivity range at the low frequency. The smoother suppresses high-mode vibrations to near-zero values for all the beam length shown in the Figure 5.7. This is also because the smoother has low-pass filtering effect.

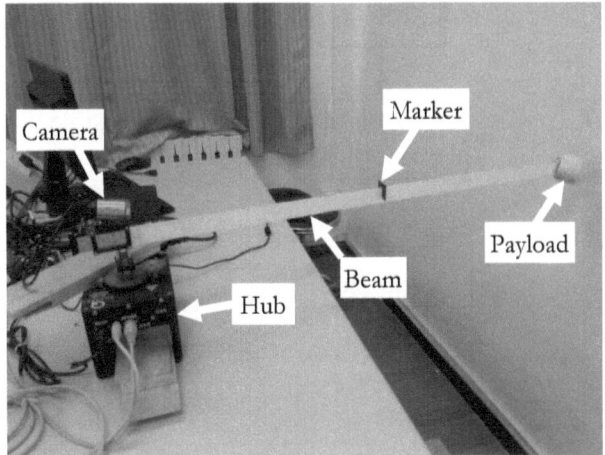

Figure 5.8 A single-link flexible manipulator.

5.1.3 Experiments

Experiments were performed on a Quanser testing apparatus shown in Figure 5.8. A motor with encoder drove the hub, to which a flexible beam is mounted. A motion control card connects amplifier to a MATLAB script in the personal computer. The length, breadth, and thickness of the beam were 950 mm, 39 mm, and 1.02 mm, respectively. A tennis ball with the mass of 121 g, which was served as the payload, connected with the beam. Vibrations of the payload have few effects on the motion of the hub because many metal blocks are mounted to the hub. The first-mode natural frequency and damping ratio in experiments were 3.35 rad/s and 0.06, respectively. The displacement of a black marker on the beam was recorded by a camera, which was also mounted to the hub. The marker in experiments was set to the middle point of the beam. Then mid-point responses based on the black marker are the experimental result.

Experimental control architecture is shown in Figure 5.9. The operator produced a bang-coast-bang acceleration (trapezoidal velocity profile) via the control interface. The original command is sent to a four-pieces smoother, which was described in the Chapter 2. Then the smoother generates a driving command for rotating the hub in the flexible manipulator. The smoother is designed using the first-mode frequency in the flexible manipulator. The design frequency, ω_m, was estimated by using mass ratio, h, and beam length, l_b.

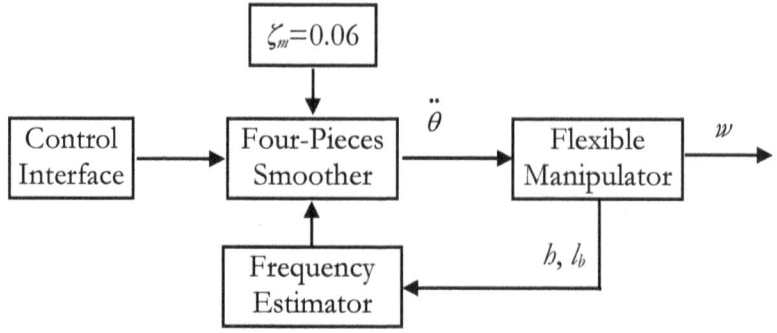

Figure 5.9 Control architecture.

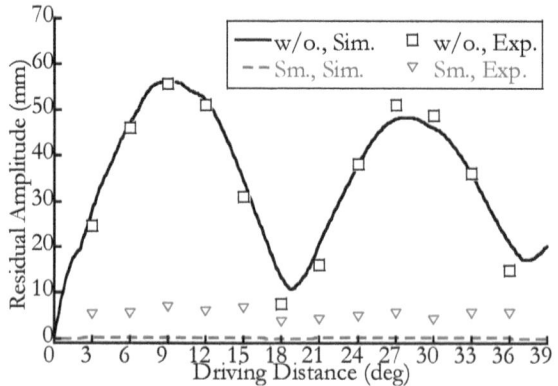

Figure 5.10 Residual amplitudes for various driving distances.

The first set of experiments investigated the effect of varying driving distances. Figure 5.10 shows simulated and experimental results of residual amplitudes. Without the controller, peaks and troughs occurred as the driving distance changed. The four-pieces smoother was designed when the design frequency and damping ratio were 3.35 rad/s and 0.06, respectively. The controlled results were limited to low levels. Experimental values were worse than simulated results because there were a small modeling error in the design frequency and uncertainty in the overall system. However, experimental data follow the general shape as the simulated curve.

The second set of experiments investigated the effect of various modeling errors in the design frequency. The experimental responses to a small modeling error in the design frequency are shown in Figure 5.11.

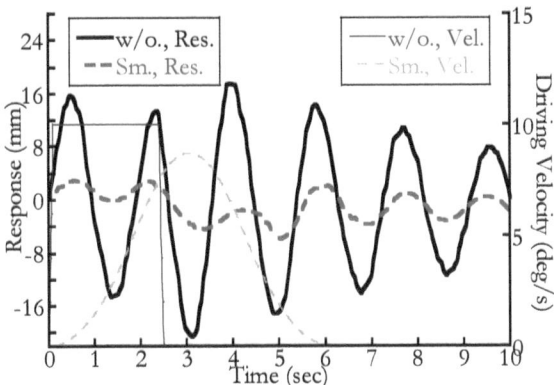

Figure 5.11 Responses to a small modeling error in the frequency.

Without the controller, transient and residual vibrations had a response with amplitudes of 30.3 mm and 38.2 mm, respectively. The residual amplitude was larger than the transient deflection because vibrations caused by the acceleration and deceleration were in phase. The four-pieces smoother was designed when the design frequency was held constant at 3.35 rad/s. The transient deflection and residual amplitude were 8.7 mm and 4.6 mm, respectively. The four-pieces smoother dramatically reduced vibrations in the case of small modeling errors.

Experimental responses for the design frequency of 6.7 rad/s are shown in Figure 5.12. Note that the design frequency of 6.7 rad/s is corresponding to the normalized frequency of 0.5. The transient deflection with the smoother was 14.4 mm, while the residual amplitude was 12.2 mm. The residual amplitude with the design frequency of 6.7 rad/s are relatively larger than that with the design frequency of 3.35 rad/s. This is because the four-pieces smoother is less insensitive at the low frequency.

Experimental responses when the design frequency was fixed at 1.675 rad/s (corresponding to the normalized frequency of 2) are shown in Figure 5.13. The transient deflection and residual amplitude were 5.7 mm and 1.7 mm, respectively. The residual amplitude was limited to a low level in this case because of the low-pass filtering effect.

Experimental results in the Figures 5.11-5.13 indicate that the four-pieces smoother provides good robustness to modeling errors in the design frequency. The experimental findings demonstrated that the four-pieces smoother is effective and robust for varying driving motions over a wide range of modeling errors in the flexible link manipulator.

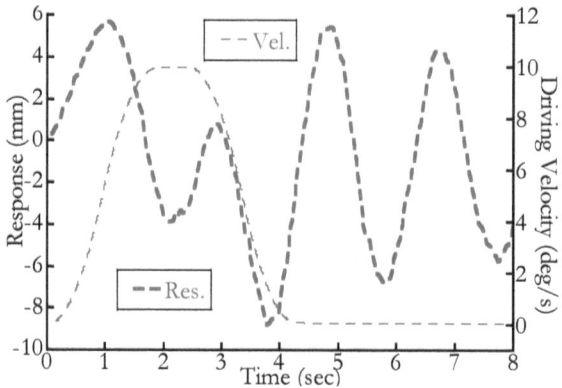

Figure 5.12 Responses to a negative modeling error in the frequency.

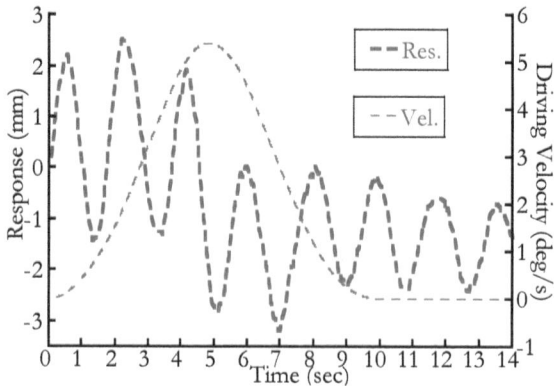

Figure 5.13 Responses to a positive modeling error in the frequency.

5.2 Duffing-Oscillator Dynamics

Duffing oscillators are widely applied in many types of mechanical systems, such as pendulums, beams, cables, and nonlinear vibration isolators. Many methods have been reported to control the Duffing oscillator, such as time delayed feedback control [110-118], combined linear-plus-nonlinear feedback control [119], state feedback control [120], optimal polynomial control [121], and sliding mode control [122]. Chen *et al.* proposed two-step and three-step shaped-command methods for controlling Duffing oscillators [123]. The magnitude and time of the shaped commands were derived using an optimization program. The effectiveness of the two- and three-step shaped-command methods was verified by simulating a fixed-fixed beam. However, the computational load for optimizing shaped commands on-the-fly is an obstacle in the application of control methods discussed in [123]. Moreover, accurately sensing vibrations is a barrier in the application of feedback controllers. Thus, it is highly desirable to control Duffing oscillators without measuring vibrations and performing heavy computational tasks.

5.2.1 Dynamics

The equation of the motion from force balance is also given in the equation (5.1). Similarly, the equation of the motion from moment balance can be written as the equation (5.2). From the mechanics of materials (flexure formula), the relationship between the bending moment and beam deflection is given by:

$$M(x,t) = EI \cdot \frac{\dfrac{\partial^2 w(x,t)}{\partial x^2}}{[1 + (\dfrac{\partial w(x,t)}{\partial x})^2]^{\frac{3}{2}}}, \tag{5.25}$$

where I is the moment of inertia of the cross section of the beam, and E is Young's modulus. Expanding the right hand term of equation (5.25) into Taylor series, and ignoring high-order terms in the resultant equation yield:

$$M(x,t) = EI \cdot \frac{\partial^2 w(x,t)}{\partial x^2} \cdot [1 - 1.5(\frac{\partial w(x,t)}{\partial x})^2]. \tag{5.26}$$

Substituting equation (5.26) into equation (5.2) and then substituting into equation (5.1), and ignoring terms involving high-order powers in dx yield:

$$EI \frac{\partial^4 w(x,t)}{\partial x^4} - 1.5EI \frac{\partial^2 (\frac{\partial^2 w(x,t)}{\partial x^2}(\frac{\partial w(x,t)}{\partial x})^2)}{\partial x^2} + \rho \frac{\partial^2 w(x,t)}{\partial t^2} = -\rho x \cdot a(t).$$

$$\tag{5.27}$$

The boundary conditions are:

$$w(x,t)|_{x=0} = 0, \tag{5.28}$$

$$\frac{\partial w(x,t)}{\partial x}|_{x=0} = 0, \tag{5.29}$$

$$\frac{\partial^2 w(x,t)}{\partial x^2}|_{x=l_b} = 0, \tag{5.30}$$

$$\frac{\partial}{\partial x}[EI \frac{\partial^2 w(x,t)}{\partial x^2}]|_{x=l_b} = m_p \frac{\partial^2 w(x,t)}{\partial t^2}|_{x=l_b}. \tag{5.31}$$

The condition (5.28) indicates that the hub base does not experience any deflection. The condition (5.29) specifies that the derivative of the deflection function at the hub base is zero. The condition (5.30) shows that there is no bending moment acting at the free end of the beam. The condition (5.31) states that no shear force acts at the free end of the beam.

Using the mode superposition method, the deflection of the beam can be expressed as a linear combination of normal modes:

$$w(x,t) = \sum_{k=1}^{+\infty} \phi_k(x) \cdot q_k(t),$$
(5.32)

The linear frequency ω_k, and the mode shape $\phi_k(x)$, can be derived by solving resultant equations (5.28-5.32):

$$\omega_k = \beta_k^2 \cdot \sqrt{\frac{EI}{\rho}},$$
(5.33)

$$\cos(\beta_k l_b)\cosh(\beta_k l_b) + 1 = h \cdot \beta_k l_b [\sin(\beta_k l_b)\cosh(\beta_k l_b) - \cos(\beta_k l_b)\sinh(\beta_k l_b)]'$$
(5.34)

$$\phi_k(x) = \sin(\beta_k x) - \sinh(\beta_k x) - r\cos(\beta_k x) + r\cosh(\beta_k x),$$
(5.35)

$$r = \frac{\sin(\beta_k l_b) + \sinh(\beta_k l_b)}{\cos(\beta_k l_b) + \cosh(\beta_k l_b)}.$$
(5.36)

where ω_k is the linear frequency of the kth mode, and h is the ratio of the payload mass to the beam mass. The numerical solution for the coefficient, β_k, will be sought by using the mass ratio, h, and beam length, l_b.

Substituting equation (5.32) into equation (5.27), and ignoring the modal coupling effects between different normal modes, result in:

$$\rho \sum_{k=1}^{+\infty} [\phi_k \frac{d^2 q_k}{dt^2}] + EI \sum_{k=1}^{+\infty} [\frac{d^4 \phi_k}{dx^4} q_k] - 1.5EI \sum_{k=1}^{+\infty} [\frac{d^2 (\frac{d^2 \phi_k}{dx^2}(\frac{d\phi_k}{dx})^2)}{dx^2} q_k^3] = -\rho x \cdot a(t)$$
(5.37)

Multiplying by $\phi_k(x)$ on both sides, and integrating over the length of the beam ($0 \leq x \leq l_b$), yield:

$$\frac{d^2 q_k}{dt^2} + \omega_k^2 q_k + e_k \omega_k^2 q_k^3 = -\gamma_k \cdot a(t); \quad \omega_k > 0,$$
(5.38)

where ω_k is the linearized natural frequency of the kth mode, and e_k is the nonlinear stiffness parameter of the kth mode. The nonlinear stiffness, e_k, and coefficient, γ_k, are determined by:

$$e_k = \frac{-1.5 \int_0^{l_b} \frac{d^2 \left(\frac{d^2 \phi_k}{dx^2} \left(\frac{d\phi_k}{dx} \right)^2 \right)}{dx^2} \phi_k dx}{\int_0^{l_b} \frac{d^4 \phi_k}{dx^4} \phi_k dx}, \qquad (5.39)$$

$$\gamma_k = \frac{\int_0^{l_b} x \phi_k dx}{\int_0^{l_b} \phi_k \phi_k dx}. \qquad (5.40)$$

Considering including a proportional damping term, equation (5.38) can be rewritten as:

$$\frac{d^2 q_k}{dt^2} + 2\zeta_k \omega_k \frac{dq_k(t)}{dt} + \omega_k^2 q_k + e_k \omega_k^2 q_k^3 = -\gamma_k \cdot a(t); \quad \omega_k > 0, \zeta_k \geq 0, \quad (5.41)$$

where ζ_k is the damping ratio of the kth mode. The model (5.41) includes an infinite number of uncoupled Duffing oscillators. Moreover, the sum of the response from each of infinite Duffing oscillators is the deflection of the beam.

Numerical simulations and experimental tests can validate the dynamics of the flexible manipulator and its corresponding equation of the motion. Numerical simulations are conducted on the first four modes of infinite Duffing oscillators for simplicity. A trapezoidal velocity profile (bang-coast-bang acceleration) drove the hub. The Young's modulus, moment of inertia, linear mass density of the beam, beam length, mass ratio, damping ratio, maximum slewing acceleration, and maximum slewing velocity used in numerical simulations are 2.06x10⁵ MPa, 3.449 mm⁴, 0.3143 kg/m, 95 cm, 0.5, 0.03, 200°/s², and 20°/s, respectively.

Figure 5.14 shows experimental and simulated responses to a trapezoidal velocity command for a slewing distance of 54 deg. The experimental data follow a similar shape pattern to the simulated curve. Figure 5.14 also gives the simulated response for the corresponding linear model, in which the nonlinear stiffness parameter is set to be $e_k=0$. With the experimental result, the nonlinear model matches better than the corresponding linear model. The comparing results indicate the correctness of the nonlinear modeling. With the experimental vibration period, the vibration period for the

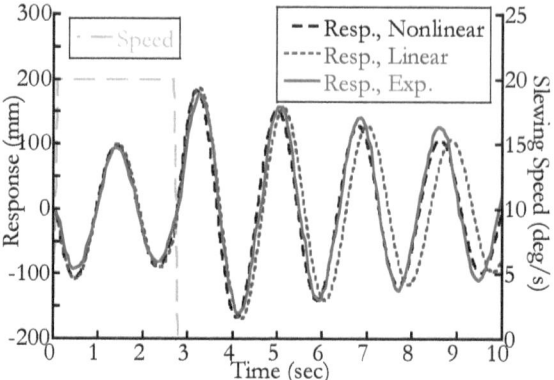

Figure 5.14 Experimental and simulated response.

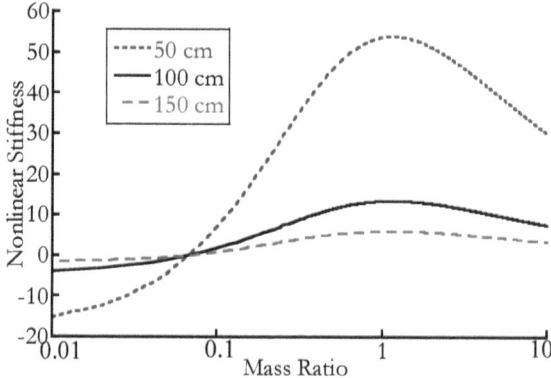

Figure 5.15 Nonlinear stiffness parameters of the first mode.

nonlinear model is smaller than that for the linear model because the nonlinear stiffness parameter is positive in this case. The hub rotates between 0s and 2.7s in Figure 5.14. The experimental result of the transient deflection and residual amplitude are found to be 186.1 mm, and 338.7 mm, respectively.

The nonlinear stiffness parameter of the first mode as a function of the mass ratio and beam length is shown in Figure 5.15. Increasing beam length decreases the magnitude of the nonlinear stiffness parameters. When the mass ratio is smaller than 0.07, the nonlinear stiffness parameter is negative, which corresponds to the softening spring type. The positive nonlinear stiffness parameter corresponds to the type of hardening spring. The nonlinear stiffness parameters increase as the mass ratio increases before 1.16. After this critical mass ratio, the nonlinear stiffness parameters decrease with an increase of the mass ratio.

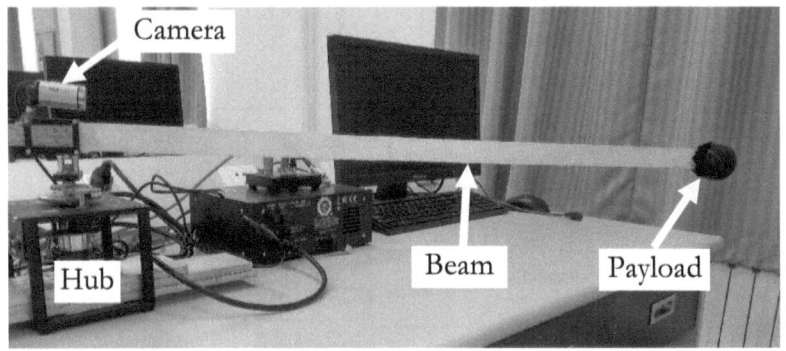

Figure 5.16 Test bench of a flexible single-link manipulator.

5.2.2 Experimental Validation

Experiments are conducted on a test rig shown in Figure 5.16. A flexible beam is mounted to a hub that is driven by a motor with encoder. The control hardware consists of a personal computer for program development and user interface, and a motion control card connecting amplifier to the personal computer. The length, width, and thickness of the beam are 950 mm, 39 mm, and 1.02 mm, respectively. A tennis ball is served as the payload, which is attached to the free end of the beam. The mass of the ball is 121 grams. A number of heavy metal blocks are mounted to the hub such that vibrations of the payload have negligible effects on the motion of the hub. A camera is mounted to the hub to record deflections of the payload. The maximum slewing velocity and maximum slewing acceleration are $20°/s$, and $200°/s^2$, respectively. Experimental measurement for the first-mode linear frequency and damping ratio is found to be approximately 3.35 rad/s and 0.03, respectively.

The experimental control architecture is shown in Figure 5.17. A bang-coast-bang acceleration (trapezoidal velocity profile) is modified by the SD function or MD smoother in Chapter 2 to produce a slewing acceleration, a, for rotating the hub. The first-mode linear frequency, ω_1, is estimated using the mass ratio, h, and beam length, l_b. Solving equations (5.39) and (5.41) yields the nonlinear stiffness parameter, e_1, and the time-dependent function, q_1, of the first mode. Then the nonlinear frequency of the first mode is then estimated for the design of the SD function and MD smoother. The damping ratio of the first mode was fixed at 0.03 in the experiment.

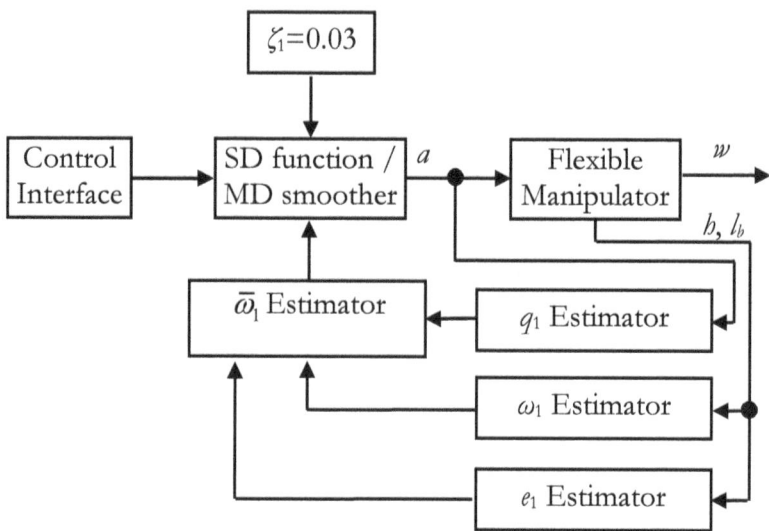

Figure 5.17 control architecture.

Both the SD function and MD smoother modify operated commands to produce shaped commands for slewing the manipulator. The comparison of experimental response under the SD function and MD smoother is shown in Figure 5.18. The transient deflections with the SD function and MD smoother are 60.4 mm and 34.8 mm, respectively. The corresponding residual amplitudes with the SD function and MD smoother are 14.1 mm and 3.7 mm, respectively. The MD smoother reduces more oscillations than the SD function. Both the SD function and MD smoother control Duffing oscillator dynamics to a very low level.

The second set of experiments investigate the effect of varying slewing distances. Figure 5.19 shows simulated and experimental results of the residual amplitude when the slewing distance is ranged from 27° to 66°. Without the controller, peaks and troughs occur as the slewing distance is changed. The experimental data match the corresponding simulated curve very well. The SD function and MD smoother were designed when the modeled linear frequency and damping ratio are 3.35 rad/s and 0.03, respectively. The SD function and MD smoother both represent a significant reduction of vibrations for all cases. The experimental data with both the SD function and MD smoother are slight higher than the simulated curve because of modeling errors in the design frequency and uncertainty in the overall system.

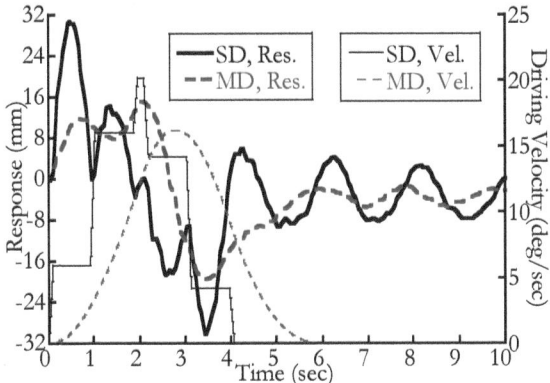

Figure 5.18 Experimental results at a slewing distance.

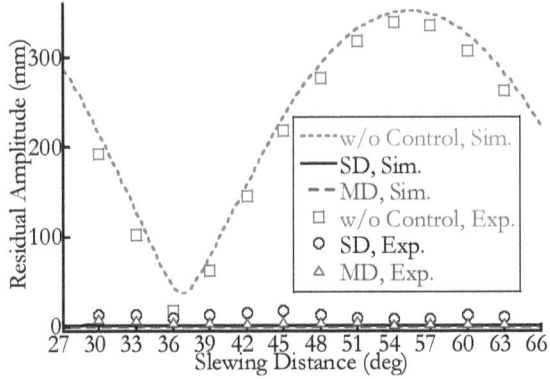

Figure 5.19 Experimental results at various slewing distances.

The last set of experiments is aimed at investigating the effect of various modeling errors in the design frequency. The experimental response to a small modeling error in the design frequency is shown in Figure 5.20. Without the controller, the transient and residual vibrations are found to have the amplitudes of 186.1 mm and 338.7 mm, respectively. The residual amplitude is larger than the transient deflection. This is because vibrations caused by the acceleration and deceleration are in phase. Both the SD function and MD smoother were designed when the designed linear frequency was held constant at 3.35 rad/s. The transient deflections with the SD function and MD smoother are 62.9 mm and 32.5 mm, respectively. Meanwhile, the residual amplitudes with the SD function and MD smoother are 8.2 mm and 2.1 mm, respectively. Both the SD function and MD smoother suppressed dramatically vibrations in the case of small modeling error. Nevertheless, residual vibrations with the MD smoother

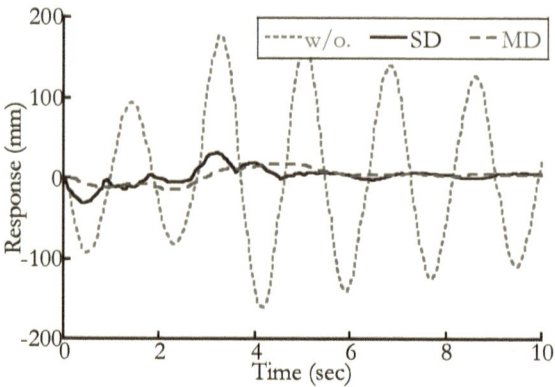

Figure 5.20 Response to a small modeling error in the frequency.

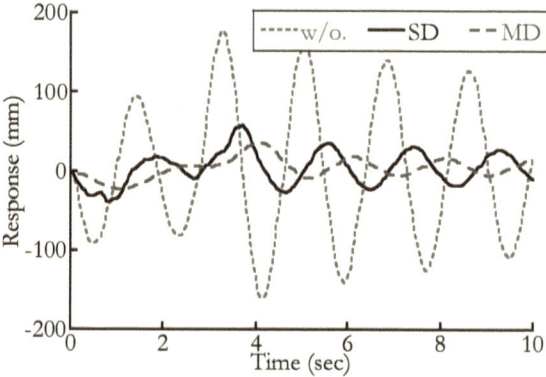

Figure 5.21 Response to a negative modeling error in the frequency.

are relatively smaller. This is because vibrations of the high-mode Duffing oscillators cannot be controlled under the SD function.

The experimental response for the modeled linear frequency of 4.79 rad/s is shown in Figure 5.21. Note that the modeled frequency of 4.79 rad/s is corresponding to -30% error. Transient deflections with the SD function and MD smoother are 156.2 mm and 58.8 mm, respectively. Meanwhile, residual amplitudes with the SD function and MD smoother are 62.3 mm and 27.4 mm, respectively. Because the MD smoother is less sensitive to negative errors, the residual amplitude with the SD function is larger than that of the MD smoother.

The experimental response is shown in Figure 5.22 when the design frequency is fixed at 2.58 rad/s, which is corresponding to +30% error. Transient deflections with the SD function and MD smoother are 85.5 mm

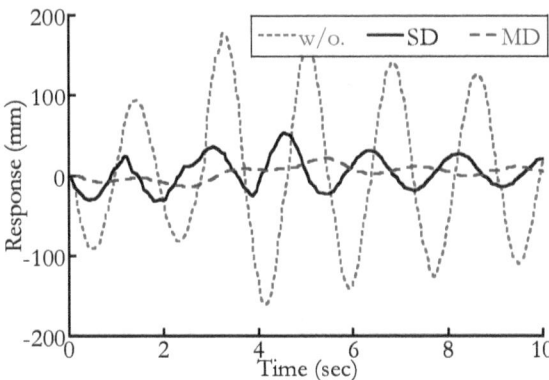

Figure 5.22 Response to a positive modeling error in the frequency.

and 34.9 mm, respectively. Meanwhile, residual amplitudes with the SD function and MD smoother are 34.8 mm and 12.7 mm, respectively. The MD-smoothed vibrations is suppressed to a lower level because of a wide range of insensitivity. The experimental findings demonstrate that both the SD function and MD smoother are effective to reduce vibrations of the flexible link manipulator having Duffing oscillator dynamics, and have good robustness in the frequency.

5.3 Chapter Summary

This chapter presented the study of the high-mode dynamics and four-pieces smoother to reduce vibrations in single-link flexible manipulators. The theoretical analyses indicate that high modes still have some impacts on the system dynamics under some specific conditions. The four-pieces smoother reduces vibrations of the infinite modes by smoothing the driving commands. Robustness analyses show that the four-pieces smoother has more insensitivity in the frequency. Numerous simulations and experiments verify that the four-pieces smoother is effective to control vibrations of the single-link flexible manipulator with infinite modes, and is robust for large range of modeling errors and working conditions.

This chapter also presented the modelling of the Duffing oscillator dynamics in flexible link manipulators, and proposed two methods to suppress vibrations of flexible single-link manipulators. The theoretical analyses indicate that flexible link manipulators can exhibit Duffing oscillator dynamics. The nonlinear frequency of the flexible manipulator increases with an increase of vibration amplitudes for large mass ratios, and decreases with increasing vibration amplitudes for small mass ratios. Robustness analyses show that the SD function and MD smoother both have more insensitivity to modeling errors in the system parameter. Numerical simulations and experiments confirmed that the developed model is appropriate for the dynamic modeling of the flexible manipulator and control methods are robust for a large range of modeling errors and working conditions.

Chapter 6. Application in Liquid Sloshing

Motions of a free liquid surface inside its container are defined as sloshing. The interaction of the sloshing dynamics with the container has unwanted effects on the overall system. Those impacts cause detrimental problems in many industrial applications ranging from packing engineering to space vehicles [124-126]. Therefore, there exists a need for a control system that can effectively reduce sloshing for safe operations.

6.1 Planar Linear Sloshing

Numerous scientists have worked to provide solutions to challenging problems posed by the sloshing dynamics. Absorbers or baffles are applied to dissipate sloshing energy [127-129]. However, the passive technique adds additional weight and complexity to the whole system. Active control method is a further solution for sloshing suppression. Venugopal and Bernstein have been proposed two active controllers. The first one used surface pressure, whereas the second controller applied a flap actuator to the surface of the fluid. Effectiveness of the controller was demonstrated by simulations [130]. The feedback control schemes measure the sloshing motion to control the container in a closed loop, such as proportional-integral-derivative control [131], sliding-mode control [132-135], H_∞ control [136-137], and Lyapunov-based feedback control [138-139]. Kurode et al. presented a nonlinear sliding-mode controller for sloshing suppression. Effectiveness of the controller was experimental verified on a moving container [133]. Reyhanoglu et al. designed a Lyapunov-based feedback controller to suppress sloshing for a Prismatic-Prismatic-Revolute robot. Effectiveness of the control law was demonstrated in simulations [138]. Grundelius et al. used optimal control and iterative learning control to reduce sloshing in the packing industry [140]. Yano et al. applied hybrid shape approach to control sloshing. Effectiveness of the controller was demonstrated in experiments [141-143]. Gandhi and Duggal stabilized sloshing in a cylindrical tank by using translational excitation as the control input. The design of the controller was based on a Lyapunov approach, and implementation of the controller applied force feedback [144].

Open loop techniques modify inputs to create prescribed motions that induce minimal sloshing including infinite impulse response filter [145], acceleration compensation [146], and input shaper [147-152]. Feddema et al. presented an infinite impulse response filter to modify the acceleration profile to produce slosh-free motions. Experiments on a FANUC S-800 robot moving a hemispherical container of water verified the performance of the method [145]. Chen et al. reduced sloshing using acceleration compensation. Experimental results on a KUKA-KR16 industrial

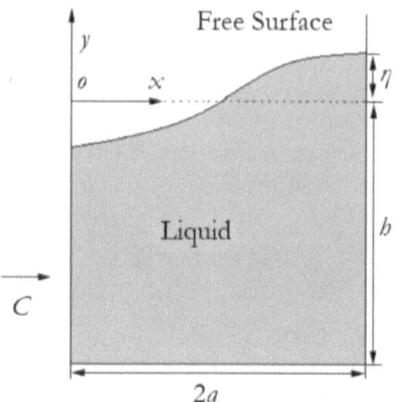

Figure 6.1 Liquid slosh in a moving container.

manipulator demonstrated the effectiveness [146]. Pridgen et al. presented a two-mode specified-insensitivity input shaper for slosh suppression. Experiments on a moving container verified the effectiveness [148]. Additionally, Significant works on the modeling of sloshing [153-156] and construction of experimental test rigs [157-158] have been reported.

6.1.1 Sloshing Dynamics

A schematic representation of sloshing in a moving container is shown in Figure 6.1. A fluid contains in a planar tank of length, $2a$. The surface of the fluid at rest is at a height, b, from the bottom of the tank. The fluid surface elevation measured from the undisturbed free surface denotes η. The acceleration of the container is $C(t)$. The following assumptions are employed to simplify the modeling of sloshing:

1). The tank is rigid and flow field is considered to be irrotational.

2). The fluid is assumed to be incompressible and nonviscous homogeneous.

3). Displacement and velocity of the liquid free surface are assumed to be small.

For the irrotational flow motion, the velocity of the fluid in the container satisfies:

$$v = v_0 + \nabla \phi, \tag{6.1}$$

where ∇ is the gradient operator, v_0 is the velocity of the tank, and ϕ is the perturbed velocity potential function. The boundary value problem is summarized as follows:

$$\nabla^2\phi = 0, \tag{6.2}$$

$$\frac{\partial\phi}{\partial x}\Big|_{x=0;2a} = 0, \tag{6.3}$$

$$\frac{\partial\phi}{\partial y}\Big|_{y=-b} = 0, \tag{6.4}$$

$$\frac{\partial\phi}{\partial y} = \frac{\partial\eta}{\partial t}, \qquad \text{at } y=\eta(x,t), \tag{6.5}$$

$$\frac{\partial\phi}{\partial t} + g\eta + C(t)x = 0, \qquad \text{at } y=\eta(x,t), \tag{6.6}$$

where g is the gravitational constant. The perturbed velocity potential function, ϕ, and surface elevation, η, satisfy:

$$\phi(x,y,t) = \sum_k \varphi_k(x,y)\dot{q}_k(t), \tag{6.7}$$

$$\eta(x,y,t) = \sum_k H_k(x,y)q_k(t), \tag{6.8}$$

where $q_k(t)$ is the time-dependent function, and $\varphi_k(x,y)$ and $H_k(x,y)$ are corresponding spatial functions. Spatial functions are the solution of following equations:

$$\nabla^2\varphi_k = 0, \tag{6.9}$$

$$\frac{\partial\varphi_k}{\partial x}\Big|_{x=0;2a} = 0, \tag{6.10}$$

$$\frac{\partial\varphi_k}{\partial y}\Big|_{y=-b} = 0, \tag{6.11}$$

$$\frac{\partial\varphi_k}{\partial y} = H_k = \frac{\omega_k^2\varphi_k}{g}, \qquad \text{at } y=\eta(x,t). \tag{6.12}$$

Solving equations (6.9-6.12) yields the natural frequency, ω_k, of sloshing modes, and spatial functions, φ_k and H_k:

$$\omega_k^2 = g \frac{k\pi}{2a} \tanh(\frac{k\pi}{2a} b) , \qquad (6.13)$$

$$\varphi_k = \cos(\frac{k\pi x}{2a}) \cosh(k\pi \frac{y+b}{2a}) , \qquad (6.14)$$

$$H_k = \frac{\omega_k^2}{g} \varphi_k , \quad k = 1, 2, 3 \ldots \qquad (6.15)$$

Equation (6.13) can be used to predict the sloshing frequency. Substituting equations (6.7) and (6.8) into equation (6.6), then multiplying by φ_k, and integrating over the free surface ($0 \leq x \leq 2a$) yield:

$$u_k \ddot{q}_k(t) + u_k \omega_k^2 q_k(t) + \alpha_k C(t) = 0 , \quad k = 1, 2, 3 \ldots \qquad (6.16)$$

where

$$u_k = \rho \int_0^{2a} H_k \varphi_k dx , \qquad (6.17)$$

$$\alpha_k = \rho \int_0^{2a} x H_k dx , \qquad (6.18)$$

, ρ is the density of the liquid. When k is even, a_k is limited to zero. Only odd sloshing modes can be excited by the horizontal acceleration $C(t)$. Equation (6.16) with the inclusion of proportional damping could be changed [159]:

$$\ddot{q}_k(t) + 2\zeta_k \omega_k \dot{q}_k(t) + \omega_k^2 q_k(t) + \frac{\alpha_k}{u_k} C(t) = 0, \quad k = 1, 3, 5 \ldots \qquad (6.19)$$

where ζ_k is the damping ratio of sloshing modes. The damping ratio is a function of the Galilei number and container geometry. The damping ratio has been developed theoretical to be approximately 0.01 for the water. Therefore, the damping ratio for each of sloshing modes is assumed to be 0.01. Substituting equation (6.15) into equation (6.8) produces the surface elevation at the measurement point:

$$\eta(x, 0, t) = \sum_k H_k(x, 0) q_k(t) = \sum_{k=odd} [\frac{\omega_k^2}{g} \varphi_k q_k(t)] . \qquad (6.20)$$

Resulting from equations (6.19) and (6.20), the transfer function of the surface elevation at the rightmost edge of the container is:

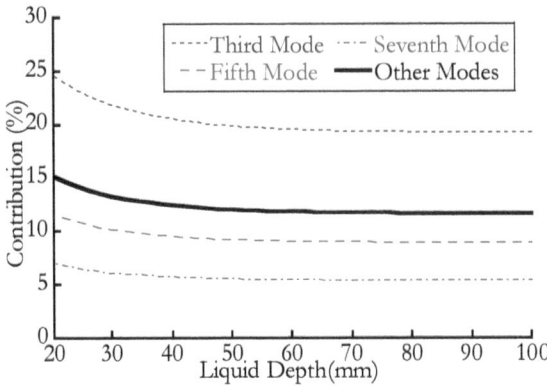

Figure 6.2 Relative amplitude contribution of high modes.

$$\eta_{x=2a,y=0}(s) = \sum_{k=odd} \left[\frac{8a}{gk^2\pi^2} \cdot \frac{-\omega_k^2}{(s^2 + 2\zeta_k\omega_k s + \omega_k^2)} C(s) \right]. \tag{6.21}$$

Then substituting equations (6.14) and (6.19) into equation (6.7) yields the transfer function of the perturbed velocity potential function at the rightmost edge of the container:

$$\phi_{x=2a,y=0}(s) = \sum_{k=odd} \left[\frac{8a}{k^2\pi^2} \cdot \frac{-s}{(s^2 + 2\zeta_k\omega_k s + \omega_k^2)} C(s) \right]. \tag{6.22}$$

The model (6.21) and (6.22) include an infinite number of sloshing modes. A sum of response for each of an infinite number of sloshing modes is the total system response. The ratio of vibration amplitude of the impulse response from the high mode to that from the first mode is defined as the relative amplitude contribution, which is applied to evaluate the effect of high sloshing modes.

The relative amplitude contribution of high modes for a large range of liquid depths is shown in Figure 6.2. The relative amplitude contribution decreases as the liquid depth increases before 50 mm, and then changes slightly as the liquid depth increases after this point. The average relative amplitude contribution of the third mode, fifth mode and seventh mode are 19.8 %, 9.2 %, 5.6 %, respectively. Meanwhile, the average relative amplitude contribution of the sum of other six higher modes is 12.0 %. Thus, theoretical results indicate that high sloshing modes have some impacts on the system dynamics. Therefore, reducing effectively sloshing induced by total modes is essential.

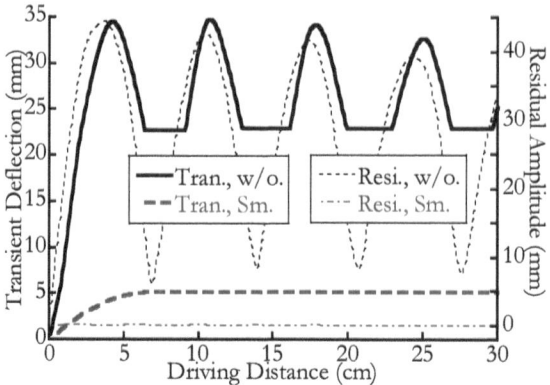

Figure 6.3 Transient and residual vibrations against driving distance.

6.1.2 Numerical Verification

Numerical verification of the two-pieces smoother over a large range of motions and liquid depths is presented in this section. A model including first four modes is used in the simulation. A trapezoidal-velocity profile drove the container. The container length, maximum velocity, and acceleration are 92 mm, 0.2 m/s, and 2 m/s², respectively. There are two stages in the system response. The peak-to-peak deflection when the container is moving is referred to as the transient deflection. The peak-to-peak deflection when the container is stopped is defined as the residual amplitude.

The transient deflection and residual amplitude of sloshing induced by various driving motions are shown in Figure 6.3. The liquid depth was set to 92 mm. Without the controller, peaks and troughs in the transient deflection and residual amplitude arise. This is because the transient deflection and residual amplitude are the result of the interference between the sloshing caused by the acceleration and deceleration. The two-pieces smoother suppresses the transient deflection and residual amplitude by an average of 82.2 % and 99.8 %, respectively. The two-pieces smoother reduces the transient and residual sloshing to a low level for various driving motions.

The transient deflection and residual amplitude of sloshing for varying liquid depth are shown in Figure 6.4. The driving distance was fixed at 20 cm. Without the controller, the driving command causes greatest transient deflection and residual amplitude at all liquid depths. The transient deflection and residual amplitude increase as the liquid depth increases before a shallow liquid depth. Then after this shallow depth, the transient

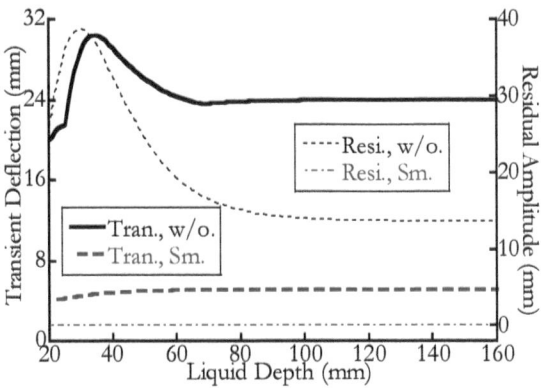

Figure 6.4 Transient and residual vibrations against liquid depth.

deflection and residual amplitude decrease with increasing liquid depth. However, the transient deflection and residual amplitude change very little when the liquid depth increases after the container length. The two-pieces smoother suppresses the transient deflection and residual amplitude by an average of 78.8 % and 99.8 %, respectively. Thus, the two-pieces smoother reduces the transient and residual sloshing to a low level over a wide range of liquid depths.

6.1.3. Experimental Verification

Figure 6.5 shows a container and camera mounted to a XY gantry, in which experiments were conducted. Panasonic AC servomotors with encoders drove the gantry. A DSP-based motion control card (Googol GT-400-SV-PCI) connects a personal computer to servo amplifier. The baseline command is sent to a Visual C++ program, which uses the two-pieces smoother algorithm, and produces the smoothed command for the drive. The sloshing frequencies were estimated by measuring the liquid depth via a plastic ruler. The travel of the gantry was 30 cm, and the size of the container was 92 mm x 92 mm x 180 mm. The gantry mounted a CMOS camera to record surface elevation at the rightmost edge of the container.

Experimental responses of the surface elevation at the rightmost edge of the container caused by a trapezoidal-velocity command is shown in Figure 6.6. The liquid depth and driving distance were set to 92 mm and 22 cm, respectively. The entire surface was level between 0 and 1 s. The container accelerated at 1 s, and decelerated at 2.1 s. Without the controller, the transient deflection and residual amplitude were 17.4 mm and 7.7 mm, respectively. The residual amplitude is smaller than the transient deflection because the sloshing induced by the acceleration is out of phase with that caused by the deceleration. With the two-pieces smoother, the transient

Figure 6.5 A moving liquid container [24].

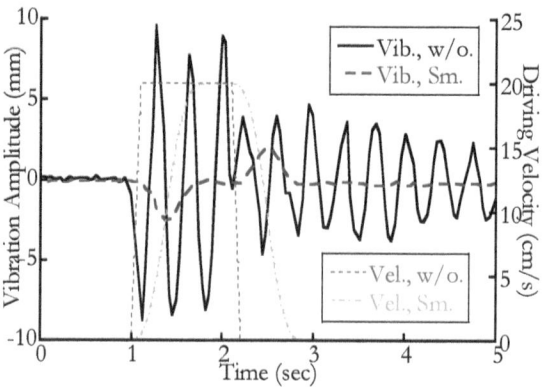

Figure 6.6 Experimental sloshing response to driving commands.

deflection and residual amplitude were 4.3 mm and 0.2 mm, respectively. It is clear that the two-pieces smoother dramatically suppresses both transient and residual sloshing.

Two sets of experiments were performed to verify the dynamics and the effectiveness of the smoother on suppressing sloshing for variations of system parameters and operation conditions. The effect of variation in the driving distance is investigated in the first set of experiments. The liquid depth was set to 92 mm. The experimental transient deflection is shown in Figure 6.7. Without the controller, the transient deflection varied with changing driving distance because of the interference between the sloshing caused by the acceleration and deceleration. A smooth velocity profile was produced by the two-pieces smoother for moving the container. The smooth transitions between boundary conditions reduced the transient sloshing. The two-pieces smoother suppressed the transient deflection by an average of 78.3 %.

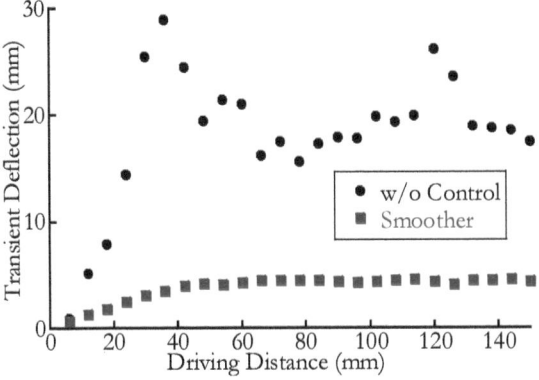

Figure 6.7 Experimental transient deflection induced by driving motions.

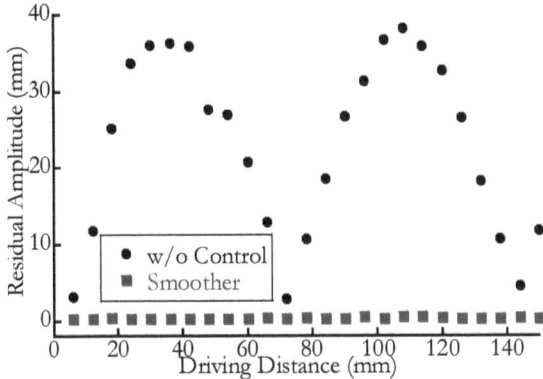

Figure 6.8 Experimental residual amplitude induced by driving motions.

Figure 6.8 shows experimental residual amplitude as a function of driving distance. Without the controller, peaks and troughs occurred as the driving distance varied. The two-pieces smoother with the notch and low-pass filtering effect suppressed the residual sloshing for infinite sloshing modes. The two-pieces smoother reduced the residual amplitude to <0.43 mm for all driving distances tested.

Effectiveness of the smoother at reducing sloshing from varying liquid depths was verified in another set of experiments. The transient deflection from these tests is shown in Figure 6.9. Without the controller, transient deflection varied slightly with changing liquid depth. The smooth velocity profile, which was produced by the two-pieces smoother, reduced the transient sloshing. The two-pieces smoother reduced the transient sloshing by an average of 79.1 %.

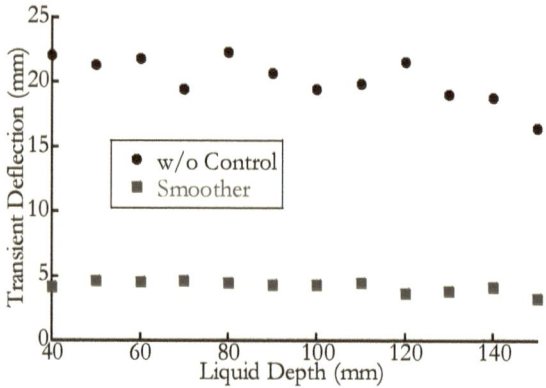

Figure 6.9 Experimental transient deflection for varying liquid depths.

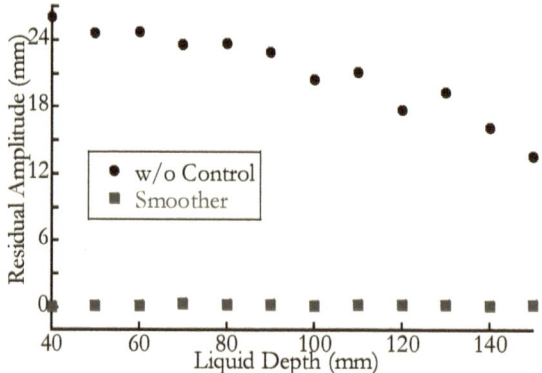

Figure 6.10 Experimental residual amplitude for varying liquid depths.

Figure 6.10 shows the residual sloshing for varying liquid depths. Without the controller, the residual amplitude decreased as liquid depth increased. The two-pieces smoother reduced the residual amplitude to <0.32 mm for all the depths tested. The notch and low-pass filtering effect of the two-pieces smoother suppressed the residual sloshing for infinite sloshing modes. Thus, the two-pieces smoother was effective for all liquid depths. These experiments demonstrated that the two-pieces smoother can effectively eliminate the transient and residual sloshing induced by various combinations of driving motions and system parameters.

6.2 Three-Dimensional Linear Sloshing

6.2.1 Sloshing Dynamics

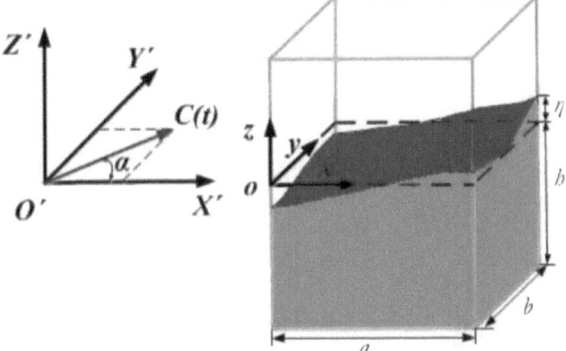

Figure 6.11. A three-dimensional sloshing model in a rectangular container.

A schematic representation of the three-dimensional sloshing in a moving rectangular liquid container is shown in Figure 6.11. A liquid fluid contains in a rectangular tank of length a, and width b. Fluid surface at rest is at height, h, from the bottom of the tank. The fluid surface elevation measured from the undisturbed free surface denotes η. Motions of the container include the linear movement of the container along the directions, X' and Y', in the inertial Newtonian frame. The acceleration, $C(t)$, of the container and steering angle, a, between the acceleration and the X' direction are defined as input variables. The moving coordinates $oxyz$ are fixed to the tank, and the oxy plane coincides with the undisturbed free surface. Axes of the moving coordinate $oxyz$ system are parallel to the inertial Newtonian frame $O'X'Y'Z'$.

The fluid is incompressible, inviscid, and irrotational. The displacement and velocity of the liquid free surface are assumed to be small. The container is rigid and impermeable. For irrotational flow motions, the absolute velocity of the fluid in the container satisfies:

$$v = v_0 + \nabla\phi, \tag{6.23}$$

where ∇ is the gradient operator, v_0 is the velocity of the tank, and ϕ is the perturbed velocity potential function. The boundary value problem in terms of the perturbed velocity potential in the moving coordinate system is given by:

$$\nabla^2\phi = 0, \tag{6.24}$$

$$\left.\frac{\partial\phi}{\partial x}\right|_{x=0,a} = 0, \tag{6.25}$$

$$\frac{\partial \phi}{\partial y}\bigg|_{y=0,b} = 0, \tag{6.26}$$

$$\frac{\partial \phi}{\partial z}\bigg|_{z=-b} = 0, \tag{6.27}$$

$$\frac{\partial \phi}{\partial z} = \frac{\partial \eta}{\partial t}, \quad \text{at } z = \eta(x, y, t), \tag{6.28}$$

$$\frac{\partial \phi}{\partial t} + g\eta + xC(t)\cos\alpha + yC(t)\sin\alpha = 0, \quad \text{at } z = \eta(x, y, t), \tag{6.29}$$

where g is the gravitational constant. The perturbed velocity potential function, ϕ, and surface elevation, η, satisfy:

$$\phi(x, y, z, t) = \sum_{ij} \varphi_{ij}(x, y, z)\dot{q}_{ij}(t), \tag{6.30}$$

$$\eta(x, y, z, t) = \sum_{ij} H_{ij}(x, y, z)q_{ij}(t), \tag{6.31}$$

where $\varphi_{ij}(x, y, z)$ and $H_{ij}(x, y, z)$ are spatial functions, i and j are nonnegative integers, and $q_{ij}(t)$ is the time-dependent function. Spatial functions are the solution of following equations:

$$\nabla^2 \varphi_{ij} = 0, \tag{6.32}$$

$$\frac{\partial \varphi_{ij}}{\partial x}\bigg|_{x=0,a} = 0, \tag{6.33}$$

$$\frac{\partial \varphi_{ij}}{\partial y}\bigg|_{y=0,b} = 0, \tag{6.34}$$

$$\frac{\partial \varphi_{ij}}{\partial z}\bigg|_{z=-b} = 0, \tag{6.35}$$

$$\frac{\partial \varphi_{ij}}{\partial z} = H_{ij} = \frac{\omega_{ij}^2 \varphi_{ij}}{g}, \quad \text{at } z = \eta(x, y, t). \tag{6.36}$$

Solving equations (6.32-6.36) yields the natural frequencies, ω_{ij}, and spatial functions, φ_{ij} and H_{ij}:

$$\omega_{ij}^2 = g\pi\sqrt{(\frac{i}{a})^2 + (\frac{j}{b})^2} \cdot \tanh[\pi h\sqrt{(\frac{i}{a})^2 + (\frac{j}{b})^2}], \tag{6.37}$$

$$\varphi_{ij} = \cos(\frac{i\pi x}{a})\cos(\frac{j\pi y}{b}) \cdot \cosh[\pi(z+b)\sqrt{(\frac{i}{a})^2 + (\frac{j}{b})^2}], \qquad (6.38)$$

$$H_{ij} = \frac{\omega_{ij}^2 \varphi_{ij}}{g}. \qquad (6.39)$$

where ω_{i0} is the frequency of the transverse mode, ω_{0j} is the frequency of the longitudinal mode, and the set $\{\omega_{ij}, i \neq 0 \ \& \ j \neq 0\}$ represents the frequency of the mixed mode. The natural frequency of the sloshing mode is dependent on the container size and liquid depth. Substituting equations (6.30) and (6.31) into equation (6.29), then multiplying by φ_{ij}, and integrating over the free surface ($0 \leq x \leq a$; $0 \leq y \leq b$) yield:

$$\lambda_{ij}\ddot{q}_{ij}(t) + \lambda_{ij}\omega_{ij}^2 q_{ij}(t) + \gamma_{ij}C(t)\cos\alpha + \beta_{ij}C(t)\sin\alpha = 0, \qquad (6.40)$$

where,

$$\lambda_{ij} = \rho\int_0^a\int_0^b H_{ij}\varphi_{ij}dxdy, \qquad (6.41)$$

$$\gamma_{ij} = \rho\int_0^a\int_0^b xH_{ij}dxdy, \qquad (6.42)$$

$$\beta_{ij} = \rho\int_0^a\int_0^b yH_{ij}dxdy. \qquad (6.43)$$

From equations (6.42) and (6.43), coefficients, γ_{ij} and β_{ij} are given by:

$$\gamma_{ij} = \begin{cases} \dfrac{-2a^2\rho\omega_{ij}^2\cosh(i\pi h/a)}{g\pi^2 i^2}, i = odd, j = 0 \\ 0, others \end{cases}, \qquad (6.44)$$

$$\beta_{ij} = \begin{cases} \dfrac{-2b^2\rho\omega_{ij}^2\cosh(j\pi h/b)}{g\pi^2 j^2}, i = 0, j = odd \\ 0, others \end{cases}. \qquad (6.45)$$

From equations (6.44) and (6.45), the transverse modal response cannot be excited by the acceleration in the y direction, the acceleration in the x direction cannot excite the longitudinal modal response, and accelerations in both the x direction and y direction cannot excite the mixed modal response. Equation (6.40) with the inclusion of the proportional damping could be changed:

$$\ddot{q}_{i0}(t) + 2\zeta_{i0}\omega_{i0}\dot{q}_{i0}(t) + \omega_{i0}^2 q_{i0}(t) + \frac{\gamma_{i0}}{\lambda_{i0}}C(t)\cos\alpha = 0, \quad i = odd, \qquad (6.46)$$

$$\ddot{q}_{0j}(t) + 2\zeta_{0j}\omega_{0j}\dot{q}_{0j}(t) + \omega_{0j}^2 q_{0j}(t) + \frac{\beta_{0j}}{\lambda_{0j}}C(t)\sin\alpha = 0, \quad j = odd, \qquad (6.47)$$

where ζ_{i0} is the damping ratio of the transverse mode, and ζ_{0j} is the damping ratio of the longitudinal mode. The damping ratio is also 0.01 for the water. By substituting equation (6.39) into equation (6.31), the surface elevation at the measurement point is:

$$\eta(x,y,0,t) = \sum_{i=odd,j=0} H_{ij}(x,y,0)q_{ij}(t) + \sum_{i=0,j=odd} H_{ij}(x,y,0)q_{ij}(t)$$
$$= \sum_{i=odd} \frac{\omega_{i0}^2 \varphi_{i0}}{g} q_{i0}(t) + \sum_{j=odd} \frac{\omega_{0j}^2 \varphi_{0j}}{g} q_{0j}(t) \qquad (6.48)$$

The transfer function of the surface elevation at the diagonal corner of the container resulting from equations (6.46-6.48) is given by:

$$\eta_{x=a,y=b,z=0}(s) = \sum_{i=odd} \frac{-4a\omega_{i0}^2 \cdot C(s)\cos\alpha}{g\pi^2 i^2 (s^2 + 2\zeta_{i0}\omega_{i0}s + \omega_{i0}^2)} + \sum_{j=odd} \frac{-4b\omega_{0j}^2 \cdot C(s)\sin\alpha}{g\pi^2 j^2 (s^2 + 2\zeta_{0j}\omega_{0j}s + \omega_{0j}^2)}$$
$$\qquad (6.49)$$

The transfer function of the perturbed velocity potential function at the diagonal corner of the container can be derived by substituting equations (6.38)(6.46)(6.47) into equation (6.30):

$$\phi_{x=a,y=b,z=0}(s) = \sum_{i=odd} \frac{-4as \cdot C(s)\cos\alpha}{\pi^2 i^2 (s^2 + 2\zeta_{i0}\omega_{i0}s + \omega_{i0}^2)}$$
$$+ \sum_{j=odd} \frac{-4bs \cdot C(s)\sin\alpha}{\pi^2 j^2 (s^2 + 2\zeta_{0j}\omega_{0j}s + \omega_{0j}^2)} \qquad (6.50)$$

The sloshing resulting from accelerations along two directions is independent. Both the surface elevation and the perturbed velocity potential are the sum of the response for each of an infinite number of sloshing modes along two directions. The amplitude of surface elevation response from an impulse command at time zero resulting from equation (6.49) is:

$$A_{x=a,y=b,z=0} = \sum_{i=odd} \frac{-4a\omega_{i0}\cos\alpha}{g\pi^2 i^2 \sqrt{1-\zeta_{i0}^2}} + \sum_{j=odd} \frac{-4b\omega_{0j}\sin\alpha}{g\pi^2 j^2 \sqrt{1-\zeta_{0j}^2}}. \qquad (6.51)$$

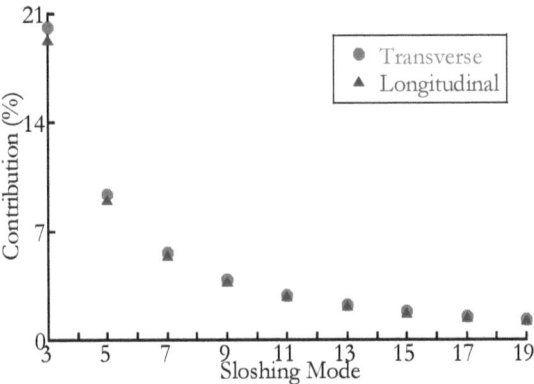

Figure 6.12 Relative amplitude contributions.

The ratio of the vibration amplitude of the impulse response from the high mode to that from the first mode is used to identify the amplitude contribution of each sloshing mode to the whole system. While the ratio is called relative amplitude contribution, c_{ij}, the equation (6.51) can be changed:

$$A_{x=a,\,y=b,\,z=0} = \frac{-4a\omega_{10}\cos\alpha}{g\pi^2\sqrt{1-\zeta_{10}^2}} \cdot \sum_{i=odd} c_{i0} + \frac{-4b\omega_{01}\sin\alpha}{g\pi^2\sqrt{1-\zeta_{01}^2}} \cdot \sum_{j=odd} c_{0j}, \qquad (6.52)$$

where ω_{10} and ω_{01} are the frequency for the first mode along the transverse and longitudinal direction, ζ_{10} and ζ_{01} are the damping ratio for the first mode along the transverse and longitudinal direction. The variables c_{i0} and c_{0j} are given by:

$$c_{i0} = \frac{\omega_{i0}\sqrt{1-\zeta_{10}^2}}{i^2\omega_{10}\sqrt{1-\zeta_{i0}^2}}, \quad i = odd, \qquad (6.53)$$

$$c_{0j} = \frac{\omega_{0j}\sqrt{1-\zeta_{01}^2}}{j^2\omega_{01}\sqrt{1-\zeta_{0j}^2}}, \quad j = odd. \qquad (6.54)$$

Thus, multiplying the sum of the relative amplitude contribution by the vibration amplitude from the first mode produces the amplitude of surface elevation responses.

The relative amplitude contribution of each sloshing modes with a liquid depth of 90 mm is illustrated in Figure 6.12. The container length, container width, maximum velocity, and acceleration are 182 mm, 102 mm, 0.2 m/s, and 2 m/s², respectively. The relative amplitude contribution of third transverse and longitudinal modes are 20.1 %, and 19.3 %, respectively. Meanwhile, the relative amplitude contribution of fifth transverse and

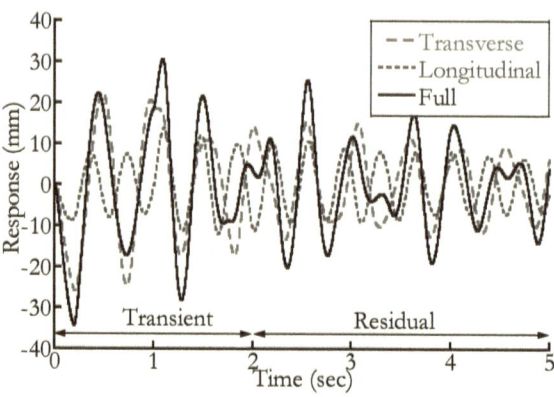

Figure 6.13 Sloshing responses resulting from two directions.

longitudinal modes are 9.4 %, and 9.0 %, respectively. In addition, the relative amplitude contribution of seventh transverse and longitudinal modes are 5.6 %, and 5.4 %, respectively. Increasing sloshing modes decreases relative amplitude contribution. However, the sum of the relative amplitude contribution from the ninth transverse mode to the nineteenth transverse mode is 13.5 %, and that from the longitudinal mode is 13.0 %. High sloshing modes have some effects on the overall system dynamics. Therefore, suppressing sloshing induced by total modes is essential.

The simulated response from first ten modes along two directions is shown in Figure 6.13. The liquid depth, driving angle, and resultant driving distance were fixed at 90 mm, 30°, and 40 cm, respectively. The transient deflection and residual amplitude of full modes are 60.8 mm, and 45.6 mm, respectively. The transient deflection and residual amplitude of the transverse mode are 48.1 mm, and 29.0 mm, respectively. Meanwhile, the transient deflection and residual amplitude of the longitudinal mode are 25.8 mm, and 21.6 mm, respectively. The sum of the response along each of two directions is the whole system response. Thus, two fundamental modes have significant effects on the sloshing dynamics.

6.2.2 Numerical Verification

Numerical robustness verification of the three-pieces smoother over a large range of motions and liquid depths is presented in this section. Using the first ten modes in each of two directions, simulations were conducted.

The transient deflection and residual amplitude induced by varying resultant driving distances are shown in Figure 6.14. The liquid depth and driving angle were fixed at 90 mm and 30°, respectively. Without the controller, the transient deflection increases as the driving distance

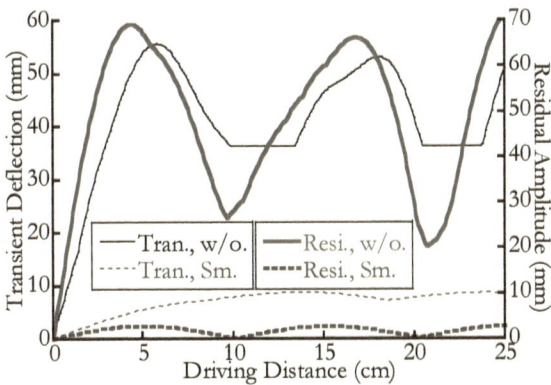

Figure 6.14 Transient and residual sloshing against driving distance.

increases before 5.8 cm. Peaks and troughs arise in the transient deflection when the sloshing induced by the acceleration and deceleration is in phase or out of phase. The residual amplitude is also the result of the interference between the sloshing caused by the acceleration and deceleration. The smoother attenuated the transient deflection and residual amplitude by an average of 83.1%, and 96.2%, respectively. The smoother created a smooth velocity profile between boundary conditions to move the container. The smooth transitions reduced the transient sloshing. The smoother suppressed the residual sloshing to a low level because the smoother is more insensitive between two modified frequencies.

The transient deflection and residual amplitude for various driving angles are shown in Figure 6.15. The resultant driving distance and liquid depth were fixed at 20 cm and 90 mm, respectively. The sloshing dynamic response is the sum of the transverse modal response and longitudinal modal response. In the case of zero driving angle, only transverse modal responses exist. Meanwhile, the container is moved longitudinally for a driving angle of 90°. Without the controller, a maximum value in the transient deflection appears near the driving angle of 48° because of the interference between the transverse and longitudinal modal response. In addition, an extreme in the residual amplitude occurs near the driving angle of 75°. With the smoother, the transient deflection and residual amplitude were suppressed by an average of 83.5 % and 98.9 %, respectively. Simulations demonstrated that the smoother has good performance at reducing the transient and residual sloshing caused by the excitation in the random direction.

The liquid depth may not be known in many cases. Therefore, it becomes important for the controller to have insensitivity to variations in the liquid depth. Figure 6.16 shows the transient deflection and residual amplitude as

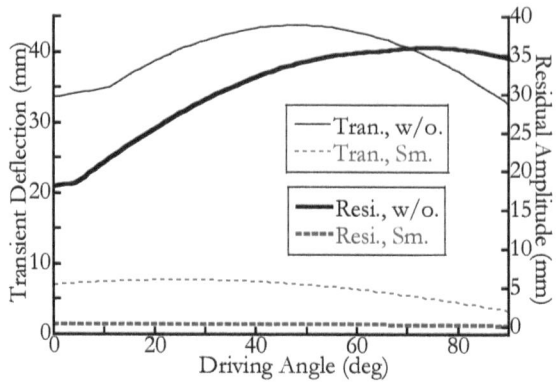

Figure 6.15 Transient and residual sloshing against driving angle.

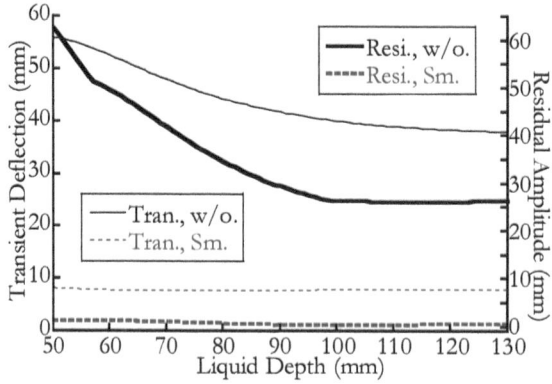

Figure 6.16 Transient and residual sloshing against liquid depth.

a function of liquid depth. The resultant driving distance and driving angle were fixed at 20 cm and 30°, respectively. Without the controller, container motions induce large sloshing. Both the transient deflection and residual amplitude decrease with increasing liquid depth, then they vary slightly as the liquid depth increases after the container width. The smoother was designed for a liquid depth of 90 mm. The transient deflection and residual amplitude were suppressed by an average of 82.4% and 98.1%, respectively. The residual amplitudes are limited to near-zero values for all liquid depths because the smoother provides more insensitivity to changes in the frequency for a wide range of liquid depths.

6.2.3 Experimental Verifications

Experiments were conducted on a testing apparatus shown in Figure 6.17. A XY gantry mounted a liquid container. The gantry is driven transversely and longitudinally by Panasonic AC servomotors with

Figure 6.17 Experimental testing apparatus.

encoders. A DSP-based motion control card connects servo amplifiers to a Visual C++ program in a personal computer. The original command is sent to the program, which applies the three-pieces smoother algorithm and produces a modified command for the drive. The size of the container was set to 182 mm x 102 mm x 200 mm. The gantry also mounts a CMOS camera for recording the surface elevation at the diagonal corner of the container.

The experimental response of the surface elevation at the diagonal corner of the container caused by the trapezoidal-velocity command is shown in Figure 6.18. The resultant distance, driving angle, and liquid depth were fixed at 20 cm, 30°, and 90 mm, respectively. The container was still between 0 and 1 s. The container accelerated at 1 s, and decelerated 1.0 s later. Without the controller, the experimental transient deflection and residual amplitude were 34.0 mm and 17.4 mm, respectively. The residual amplitude was smaller than the transient deflection. This is because the sloshing caused by the deceleration attenuated that induced by the acceleration. With the three-pieces smoother, the experimental transient deflection and residual amplitude were 5.4 mm and 0.4 mm, respectively. Experimental results demonstrated that the smoother can suppress the transient and residual sloshing.

One set of experiments were performed to verify the effectiveness of the smoother on suppressing sloshing for variations of the liquid depth. The container was driven at a distance of 20 cm while the driving angle was set to 30°. Simulated and experimental results of the transient and residual sloshing from those tests are shown in Figure 6.19. Without the controller, both the transient and residual sloshing decreased slowly as the liquid depth increased. The simulated results were worse than the experimental data

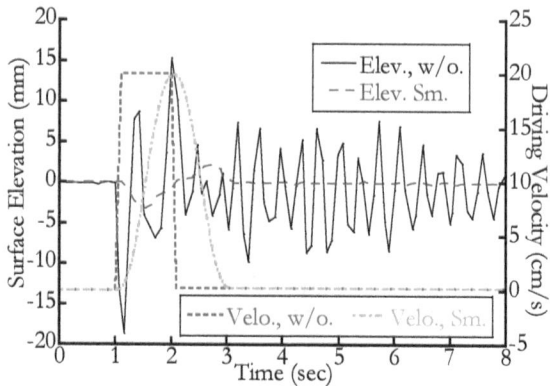

Figure 6.18 Experimental sloshing response to a driving command.

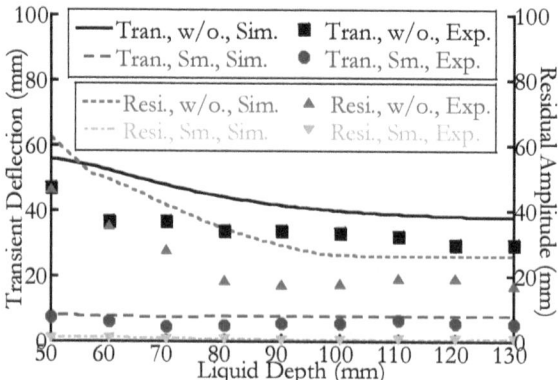

Figure 6.19 Experimental sloshing for various liquid depths.

because the liquid was assumed to be inviscid and the surface tension was ignored. The smoother was designed for a liquid depth of 90 mm. The smoother suppressed the experimental transient sloshing by an average of 83.7 %, and reduced the residual sloshing to < 0.92 mm for all depths tested. The smoother has good performance to reduce the transient sloshing because the smooth velocity profile produced by the smoother benefits reduction of the transient sloshing. Meanwhile, the smoother limited the residual sloshing to a low level. This is because the smoother is more insensitive to changes in the sloshing frequency. Those experiments verified that the smoother can effectively reduce the three-dimensional sloshing for a large range of various system parameters as it was predicted by the simulation.

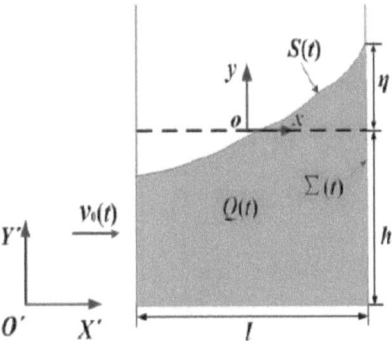

Figure 6.20 Planar nonlinear slosh model in a moving container.

6.3 Planar Nonlinear Sloshing

Significant attentions for modeling of the sloshing have been focused on the equivalent mechanical model or linear sloshing model. The planar surface without rotation of its nodal diameter (for the case of the cylindrical container) or nodal line (for the rectangular container) was kept in the equivalent mechanical model or linear sloshing model. Furthermore, the free surface of the nonlinear sloshing exhibits non-planar motions with rotation of its nodal diameter or nodal line. This nonlinear sloshing dynamics will present large amplitude oscillations. Faltinsen et al. proposed the modeling of the nonlinear sloshing [160]. Nevertheless, coefficients presented by Faltinsen are only expressed in a non-expanded fashion for three-dimensional nonlinear slosh. Meanwhile, much of work for slosh suppression has been directed at controlling the sloshing based on the equivalent mechanical model or linear sloshing model. However, no effects have been directed at controlling the nonlinear sloshing dynamics.

6.3.1 Sloshing Dynamics

A moving liquid container of length, l, partly filled by an inviscid incompressible fluid with a depth of, h, is shown in Figure 6.20. The flow is also assumed to be irrotational and the container is rigid. Let $O'X'Y'$ be the inertial coordinates and oxy be the moving coordinates fixed to the container. Note that the axes of the two coordinates are parallel to each other. The origin of the moving coordinates is in the mean free surface at the centreplane of the tank. The container was driven along the X' direction with the velocity of $v_0(t)$. $\eta(x,t)$ denotes the liquid free surface elevation measured from the undisturbed free surface.

The surface elevation, $\eta(x,t)$, and the relative velocity potential, $\varphi(x,t)$, can be expressed as:

$$\eta(x,t) = \sum_i \cos[\pi i(x/l + 0.5)] \cdot \beta_i(t), \tag{6.55}$$

$$\varphi(x,t) = \sum_i \cos[\pi i(x/l + 0.5)] \cdot R_i(t), \tag{6.56}$$

where $\beta_i(t)$ and $R_i(t)$ are time-dependent functions. The infinite dimensional mode system in (6.55) and (6.56) can be detuned to a finite-dimensional mode system by the asymptotic relation.

$$O(\beta_1) = \varepsilon^{1/3}, \quad O(\beta_2) = \varepsilon^{2/3}, \quad O(\beta_3) = \varepsilon, \tag{6.57}$$

where ε is the small parameter. Higher-orders than ε will be neglected in the nonlinear equations. Thus, the functions $\beta_i(t)$ and $R_i(t)$ are given by:

$$
\begin{aligned}
&\ddot{\beta}_1 + \omega_1^2 \beta_1 + (E_1 + \frac{2E_0}{E_1})(\ddot{\beta}_1 \beta_2 + \dot{\beta}_1 \dot{\beta}_2) + (E_1 - \frac{2E_0}{E_2})\ddot{\beta}_2 \beta_1 \\
&+ (\frac{8E_0^2}{E_1 E_2} - 2E_0)(\ddot{\beta}_1 \beta_1^2 + \dot{\beta}_1^2 \beta_1) - \frac{8E_1 l}{\pi^2}\dot{v}_0(t) = 0
\end{aligned}
\tag{6.58}
$$

$$\ddot{\beta}_2 + \omega_2^2 \beta_2 + (2E_2 - \frac{4E_0}{E_1})\ddot{\beta}_1 \beta_1 - (\frac{4E_0}{E_1} + \frac{E_2(2E_0 - E_1^2)}{E_1^2})\dot{\beta}_1^2 = 0, \tag{6.59}$$

$$
\begin{aligned}
&\ddot{\beta}_3 + \omega_3^2 \beta_3 + (3E_3 - \frac{6E_0}{E_1})\ddot{\beta}_1 \beta_2 + (\frac{24E_0^2}{E_1 E_2} - \frac{9E_0 E_3}{E_1} - 3E_0)\ddot{\beta}_1 \beta_1^2 \\
&+ (3E_3 - \frac{6E_0}{E_2})\ddot{\beta}_2 \beta_1 + (3E_3 - \frac{6E_0}{E_1} - \frac{6E_0}{E_2} - \frac{6E_0 E_3}{E_1 E_2})\dot{\beta}_1 \dot{\beta}_2 \\
&+ (\frac{48E_0^2}{E_1 E_2} + \frac{24E_0^2 E_3}{E_1^2 E_2} - \frac{24E_0 E_3}{E_1} - 6E_0)\dot{\beta}_1^2 \beta_1 - \frac{8 l E_3}{3\pi^2}\dot{v}_0(t) = 0
\end{aligned}
\tag{6.60}
$$

$$R_1 = \frac{\dot{\beta}_1}{2E_1} + \frac{E_0}{E_1^2}\dot{\beta}_1 \beta_2 - \frac{E_0}{E_1 E_2}\dot{\beta}_2 \beta_1 + \frac{E_0}{E_1}(\frac{4E_0}{E_1 E_2} - \frac{1}{2})\dot{\beta}_1 \beta_1^2, \tag{6.61}$$

$$R_2 = \frac{1}{4E_2}(\dot{\beta}_2 - \frac{4E_0}{E_1}\dot{\beta}_1 \beta_1), \tag{6.62}$$

$$R_3 = \frac{\dot{\beta}_3}{6E_3} - \frac{E_0}{E_1 E_3}\dot{\beta}_1 \beta_2 - \frac{E_0}{E_2 E_3}\dot{\beta}_2 \beta_1 + (\frac{4E_0^2}{E_1 E_2 E_3} - \frac{E_0}{2E_3})\dot{\beta}_1 \beta_1^2, \tag{6.63}$$

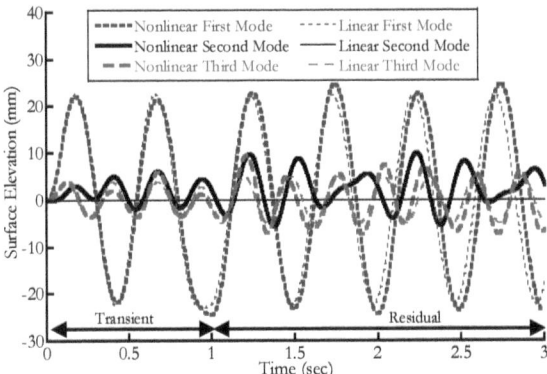

Figure 6.21 Simulated slosh response for the nonlinear and linear models.

where ω_i is the natural frequency of ith sloshing mode, and the coefficients E_i are given by:

$$E_0 = \frac{1}{8}(\frac{\pi}{l})^2; \quad E_i = \frac{\pi}{2l}\tanh(\frac{\pi i}{l}h), \quad i \geq 1.$$ (6.64)

The natural frequency of ith sloshing mode, ω_i, in the nonlinear slosh models is given by:

$$\omega_i = \sqrt{2igE_i} .$$ (6.65)

Figure 6.21 shows the simulated responses of the first three modes for both the nonlinear and linear slosh models when the driving distance, container length, and liquid depth were fixed at 22.5 cm, 182 mm, and 120 mm, respectively. The original command used to driving the container is a trapezoidal-velocity profile (bang-coast-bang acceleration). An acceleration pulse causes a rise in the velocity until the maximum velocity is reached. The container then moves at its maximum velocity until a deceleration pulse acts. This causes a decrease in the velocity until zero.

The slosh responses of the first and third modes for the nonlinear model matches that for the linear model very well. However, there is a huge difference for the second mode. The second-mode response for the linear model are limited to zero because the container motion cannot excite the even modes for the linear slosh model. Meanwhile, the second-mode response for the nonlinear model exhibits relatively large amplitude oscillations.

There are two stages in the system response. The transient stage is defined as the time frame when the container is in motion. The maximum peak-to-peak deflection during the transient stage is referred to as the

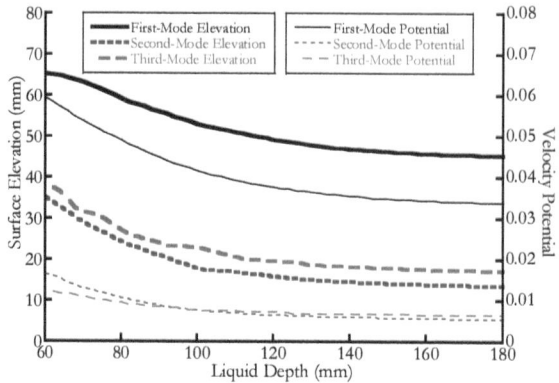

Figure 6.22 Simulated residual amplitude for the nonlinear model.

transient deflection. The residual stage is defined as the time frame when the container is stopped. The maximum peak-to-peak deflection during the residual stage is defined as the residual amplitude. Maximum driving velocity and acceleration were 0.25 m/s and 2.5 m/s², respectively. The damping ratio was zero in the simulations. Note that the duration of the simulation in this section is 5 s and the slosh measurement position is chosen at the area of the free surface near the left side of the container.

The simulated residual amplitudes of the first three sloshing modes for the nonlinear model are shown in Figure 6.22. While the average residual amplitude of the first mode and the third mode are 51.8 mm and 22.1 mm, respectively, that of the second mode reaches 18.5 mm. The maximum peak-to-peak amplitudes of the velocity potential from the first three nonlinear modes are also given in Figure 6.22. The amplitudes of the velocity potential for the second-mode slosh are larger than that for the third-mode slosh when the liquid depths are shallow. Simulated results indicate that the odd modes have large impacts on the weakly nonlinear sloshing dynamics, but the even modes may also have some effects, which is significant different from the linear slosh. Thus, there is a need for a control system that can effectively reduce weakly nonlinear slosh induced by both odd and even modes.

The higher-modes than three, which are neglected in nonlinear model, can be described in the linear model. Therefore, the slosh model in this section includes the first three nonlinear modes and other linear higher modes. The sloshing response is the sum of the response for the total modes. Simulations were conducted using the first ten modes model in this section.

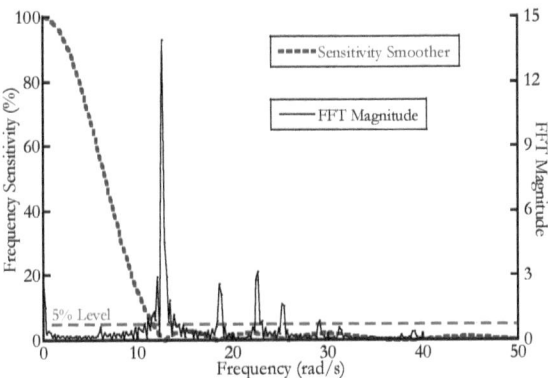

Figure 6.23 FFT magnitude of the slosh response and frequency sensitivity curve for the combined smoother.

Two one-piece smoothers with the first-mode frequency and the second-mode frequency are convolved together to create a combined smoother. The combined smoother provides notch and low-pass filtering effect. To better identify the slosh suppression, the fast Fourier transform (FFT) of a slosh response is shown in Figure 6.23 when the driving distance, container length, and liquid depth were fixed at 22.5 cm, 182 mm, and 120 mm, respectively. The estimates of the sloshing frequencies from equation (6.65) are revealed by the peaks of the FFT magnitude at 12.8 rad/s, 18.4 rad/s, and 22.5 rad/s, etc. Using the frequencies of the first two sloshing modes (12.8 rad/s and 18.4 rad/s) as design frequencies, the combined smoother would suppress slosh induced by total modes. Figure 6.23 also shows the frequency sensitive curve for the combined smoother. The frequency insensitivity, defined as the range of each curve that lies below 5% of percentage residual amplitude, provides a quantitative measure of robustness. It is clear that the combined smoother produces a low-pass filtering effect. The 5% insensitivity of the combined smoother ranges from 11.55 rad/s to infinity. Thus, the presented combined smoother could suppress a wide range of sloshing frequencies.

To better evaluate slosh suppression for high modes, the maximum peak-to-peak amplitude of the velocity potential from the first ten modes is given in Table 6.1. Without the controller, the amplitude of the velocity potential for the first nonlinear mode is larger than that for the others. Thus, the first mode is fundamental for sloshing dynamics. The maximum peak-to-peak amplitude of the second nonlinear mode is relatively large. Thus, it cannot be ignored. The velocity potential for the case of the fourth, the sixth, the eighth, and the tenth mode are limited to zero. This is because the container motion cannot trigger the even modes for the linear slosh.

Table 6.1 Velocity potential from the first ten modes

Mode	Without Control	Combined Smoother
1	3.76e-2	2.68e-5
2	6.35e-3	6.91e-5
3	7.06e-3	8.83e-5
4	0	0
5	1.01e-3	2.07e-5
6	0	0
7	1.84e-4	1.25e-6
8	0	0
9	4.15e-4	3.30e-7
10	0	0

The velocity potential resulting from the combined smoother is also given in the Table 6.1. The combined smoother reduces the velocity potential for the total modes below a very low level.

6.3.2 Simulated Results

In many cases, the liquid depth and tank length may not be known accurately and the driving distance often changes. Then it becomes important for the combined smoother to have insensitivity to the variations in the system parameters and to provide good performance over a large range of container motions.

Figure 6.24 shows the transient deflection and residual amplitude of the surface elevation induced by varying driving distances when the container length and liquid depth were fixed at 182 mm and 120 mm, respectively. The uncontrolled transient deflection increases with increasing driving distance before 9.5 cm because the amount of transient deflection depends on the size of the acceleration pulse and the duration of the transient stage. After this point, the transient deflection is dependent on the interference between the slosh caused by the acceleration and deceleration. When the vibration induced by the deceleration is in phase with that caused by the acceleration, the peaks will arise in the uncontrolled transient deflection. When the vibrations caused by the acceleration and deceleration are out of phase, the troughs occur. The combined smoother attenuated the transient deflection by an average of 78.7 %. This is because the smoother produces a smooth velocity profile to move the container. The smooth transitions between boundary conditions reduce the transient slosh. The uncontrolled residual amplitude is also the result of the interference between the slosh caused by the acceleration and deceleration. The combined smoother

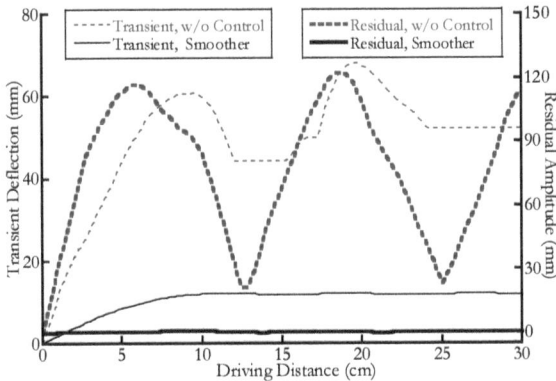

Figure 6.24 Transient and residual amplitude against driving distance.

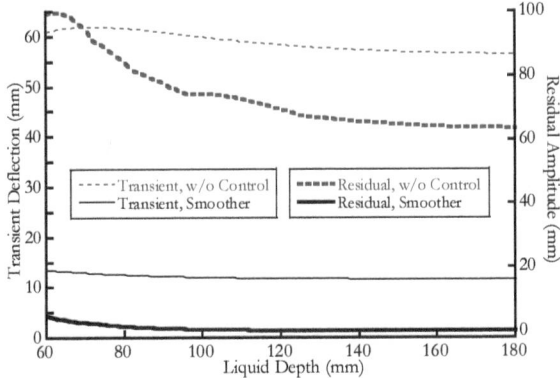

Figure 6.25 Transient and residual amplitude against liquid depth.

eliminated the residual amplitude by an average of 99.3 % because the modeled sloshing frequencies was correct.

Figure 6.25 shows the transient deflection and residual amplitude of the surface elevation as a function of liquid depth, h, when the combined smoother was designed for a liquid depth of 120 mm. The driving distance and container length were fixed at 22.5 cm and 182 mm in this case. Without the controller, the container motion results in large slosh. The uncontrolled transient deflection changes slightly with increasing liquid depth. The uncontrolled residual amplitude decreases sharply as the liquid depth increases. The combined smoother suppressed the transient deflection and residual amplitude by an average of 79.6 % and 98.7 %, respectively. The transient deflection with the combined smoother keeps low values because of the transient-vibration constraint of the smoother. The residual amplitudes with the combined smoother are limited below a

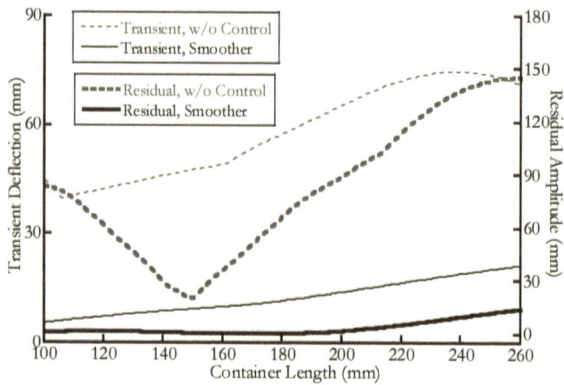

Figure 6.26 Transient and residual amplitude against container length.

low level because the combined smoother can provide more insensitivity to the changes in the frequency.

Figure 6.26 shows the transient deflection and residual amplitude of the surface elevation for various container lengths when the driving distance and liquid depth were fixed at 22.5 cm and 120 mm in this case. The transient deflection without the controller decreases before 106 mm, and then increases as the container length increases. There is a local maximum in the uncontrolled transient deflection for the container length of 239 mm. The uncontrolled residual amplitude decreases before 150 mm, then increases with increasing the container length. The trough arises at the tank length of 150 mm because the vibrations caused by the container motion are out of phase. The combined smoother was designed for a container length of 182 mm. The transient deflection with the smoother increases with increasing the container length. The combined smoother attenuated the transient deflection by an average of 78.5 %. This is because the smoother limits the transient vibrations to low level. In addition, the combined smoother eliminated that by an average of 95.7 %. This is also because the combined smoother can provide more insensitivity to the changes in the frequency.

The abovementioned simulations demonstrated that the smoother can robustly suppress both the transient and residual slosh for varying working conditions and system parameters.

6.3.3 Experimental Results

Experiments were also conducted on a testing apparatus shown in Figure 6.17. Figure 6.27 shows the experimental response of surface elevation at the left side of the container caused by the trapezoidal-velocity commands.

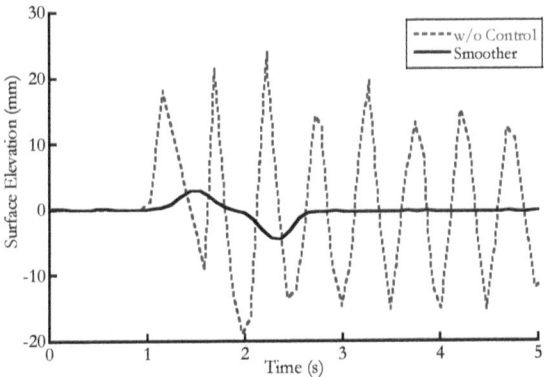

Figure 6.27 Experimental response to driving commands.

The tank was driven for a distance of 22.5 cm and the liquid depth was fixed at 120 mm. The surface was level and not moving between 0 and 1 s. The tank accelerated at 1 s, which resulted in slosh, and the tank decelerated at 0.9 s later, which resulted in additional slosh. Without the controller, the transient and residual slosh had a response with amplitudes of 40.3 mm and 38.7 mm, respectively. The uncontrolled residual amplitude was smaller than the uncontrolled transient deflection because the slosh caused by the acceleration was out of phase with that induced by the deceleration. The effect of the smoother is also shown in Figure 6.27. Experimental results show that the transient deflection and residual amplitude with the combined smoother were 7.4 mm and 0.2 mm, respectively. It demonstrated that the smoother can reduce the transient and residual slosh. This is because the combined smoother suppresses a wider range of slosh frequencies.

To verify the dynamics and effectiveness of the combined smoother on suppressing transient and residual slosh for variations of container motions and system parameters, two sets of experiments were performed. The first experiment investigated the effects of changes in the driving distance when the liquid depth was fixed at 120 mm. Figure 6.28 shows the experimental results of transient deflection and residual amplitude from these tests. Without the controller, both the transient deflection and residual amplitude varied from driving distances because of the interference between the slosh caused by the acceleration and deceleration. The combined smoother attenuated the transient deflection by an average of 79.0 %. The transient deflection with the combined smoother was limited to a lower level. In addition, the combined smoother eliminated the residual amplitude by an average of 99.3 %. The combined smoother suppressed the residual slosh to a lower level because it has better frequency insensitivity.

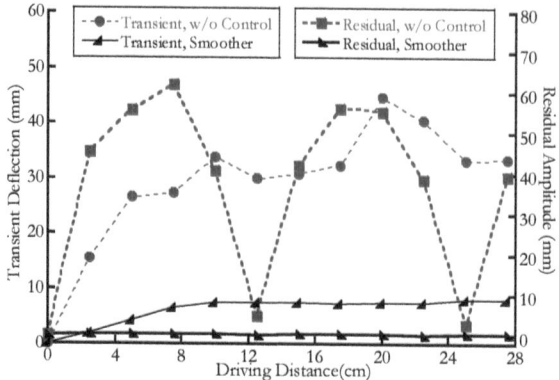

Figure 6.28 Transient and residual amplitude induced by driving motions.

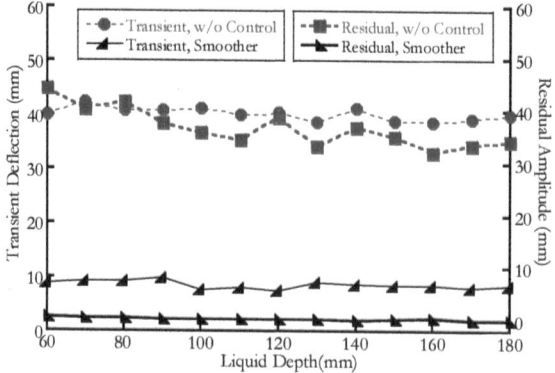

Figure 6.29 Transient and residual amplitude for various liquid depths.

Another experiment was conducted to verify the robustness of the combined smoother at suppressing transient and residual slosh for various liquid depths, h. The driving distance was fixed at 22.5 cm. The liquid depth ranged from 60 mm to 180 mm while the combined smoother was designed for a liquid depth of 120 mm. Figure 6.29 shows experimental transient deflection and residual amplitude from these tests. Similar to the simulated results, the uncontrolled transient deflection varied slightly as the liquid depth increased, and the uncontrolled residual amplitude decreased with increasing the liquid depth. Experimental results show that the combined smoother attenuated the transient deflection by an average of 78.9 %. The combined smoother provided better performance at reducing transient slosh. Meanwhile, the combined smoother suppressed the residual amplitude to < 0.67 mm for all depths tested. The combined smoother limited the residual slosh to a lower level because the combined smoother is more insensitive to changes in sloshing frequency.

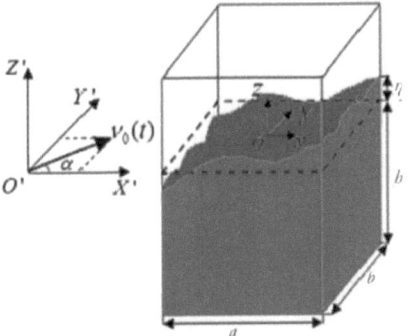

Figure 6.30 A slosh model in a rectangular tank.

Those experiments verified that the presented combined smoother can effectively suppress weakly nonlinear slosh for a wide range of various system parameters and motion conditions as it was predicted by the simulations.

6.4 Three-Dimensional Nonlinear Sloshing

6.4.1 Sloshing Dynamics

A schematic representation of three-dimensional nonlinear sloshing in a moving liquid container is shown in Figure 6.30. A rectangular tank of length, a, and breadth, b, contains an inviscid incompressible fluid. The depth of fluid is denoted as h. The container is rigid, and the sloshing does not affect motions of the container.

Let $O'X'Y'Z'$ be the inertial coordinates, while $oxyz$ be the moving coordinates. The two coordinates are parallel to each other. The mean free surface at the centreplane of the tank is the origin of the moving coordinates. The liquid free surface elevation measured from the undisturbed free surface is defined as $\eta(x,y,z,t)$. Planar excitations drive the container. The motion is divided into the linear movement of the container along two directions, X' and Y'. The motion can also be defined as: the velocity, $v_0(t)$, of the container and steering angle, a, between the velocity and the X' direction.

The equation of the surface elevation for the nonlinear sloshing is given by:

$$\eta(x,y,0,t) = a\sum_{i}\sum_{j}\cos[\pi i(\frac{x}{a}+0.5)]\cdot\cos[\pi j(\frac{ry}{a}+0.5)]\cdot\beta_{i,j}(t), \quad (6.66)$$

where $\beta_{i,j}(t)$ is the time-dependent function, and r is the ratio of the tank length to tank breadth. In this nonlinear sloshing model, higher-order modes will be neglected. Then, the time-dependent function, $\beta_{i,j}(t)$, in [160] including the proportional damping terms are given by [161]:

$$\ddot{\beta}_{1,0} + 2\zeta_{1,0}\omega_{1,0}\dot{\beta}_{1,0} + \omega_{1,0}{}^2\beta_{1,0} + d_{1,1}(\ddot{\beta}_{1,0}\beta_{2,0} + \dot{\beta}_{1,0}\dot{\beta}_{2,0})$$
$$+d_{1,2}(\ddot{\beta}_{1,0}\beta_{1,0}{}^2 + (\dot{\beta}_{1,0})^2\beta_{1,0}) + d_{1,3}\ddot{\beta}_{2,0}\beta_{1,0} + d_{1,4}\ddot{\beta}_{1,0}\beta_{0,1}{}^2$$
$$+d_{1,5}\ddot{\beta}_{0,1}\beta_{1,1} + d_{1,6}\beta_{1,0}\beta_{0,1}\ddot{\beta}_{0,1} + d_{1,7}\beta_{0,1}\ddot{\beta}_{1,1} + d_{1,8}\dot{\beta}_{1,0}\beta_{0,1}\dot{\beta}_{0,1} \quad (6.67)$$
$$+d_{1,9}\beta_{1,0}\dot{\beta}_{0,1}{}^2 + d_{1,10}\dot{\beta}_{0,1}\dot{\beta}_{1,1} + d_{1,11}v_0(t)\cos(\alpha) = 0$$

$$\ddot{\beta}_{0,1} + 2\zeta_{0,1}\omega_{0,1}\dot{\beta}_{0,1} + \omega_{0,1}{}^2\beta_{0,1} + d_{2,1}(\ddot{\beta}_{0,1}\beta_{0,2} + \dot{\beta}_{0,1}\dot{\beta}_{0,2})$$
$$+d_{2,2}(\ddot{\beta}_{0,1}\beta_{0,1}{}^2 + \dot{\beta}_{0,1}{}^2\beta_{0,1}) + d_{2,3}\ddot{\beta}_{0,2}\beta_{0,1} + d_{2,4}\ddot{\beta}_{0,1}\beta_{1,0}{}^2$$
$$+d_{2,5}\ddot{\beta}_{1,0}\beta_{1,1} + d_{2,6}\beta_{1,0}\beta_{0,1}\ddot{\beta}_{1,0} + d_{2,7}\beta_{0,1}\dot{\beta}_{1,0}{}^2 + d_{2,8}\beta_{1,0}\ddot{\beta}_{1,1} \quad (6.68)$$
$$+d_{2,9}\dot{\beta}_{1,0}\dot{\beta}_{0,1}\beta_{1,0} + d_{2,10}\dot{\beta}_{1,0}\dot{\beta}_{1,1} + d_{2,11}v_0(t)\sin(\alpha) = 0$$

$$\ddot{\beta}_{2,0} + 2\zeta_{2,0}\omega_{2,0}\dot{\beta}_{2,0} + \omega_{2,0}^2\beta_{2,0} + d_{3,1}\ddot{\beta}_{1,0}\beta_{1,0} + d_{3,2}\dot{\beta}_{1,0}{}^2 = 0 \quad (6.69)$$

$$\ddot{\beta}_{0,2} + 2\zeta_{0,2}\omega_{0,2}\dot{\beta}_{0,2} + \omega_{0,2}^2\beta_{0,2} + d_{4,1}\ddot{\beta}_{0,1}\beta_{0,1} + d_{4,2}\dot{\beta}_{0,1}{}^2 = 0, \quad (6.70)$$

$$\ddot{\beta}_{1,1} + 2\zeta_{1,1}\omega_{1,1}\dot{\beta}_{1,1} + \omega_{1,1}^2\beta_{1,1}$$
$$+d_{5,1}\ddot{\beta}_{1,0}\beta_{0,1} + d_{5,2}\ddot{\beta}_{0,1}\beta_{1,0} + d_{5,3}\dot{\beta}_{1,0}\dot{\beta}_{0,1} = 0 \quad (6.71)$$

$$\ddot{\beta}_{2,1} + 2\zeta_{2,1}\omega_{2,1}\dot{\beta}_{2,1} + \omega_{2,1}^2\beta_{2,1} + d_{6,1}\beta_{1,0}\beta_{0,1}\ddot{\beta}_{1,0} + d_{6,2}\ddot{\beta}_{1,0}\beta_{1,1}$$
$$+d_{6,3}\beta_{0,1}(\dot{\beta}_{1,0})^2 + d_{6,4}\ddot{\beta}_{0,1}\beta_{2,0} + d_{6,5}\dot{\beta}_{1,0}\dot{\beta}_{0,1}\beta_{1,0} + d_{6,6}\ddot{\beta}_{0,1}\beta_{1,0}{}^2, \quad (6.72)$$
$$+d_{6,7}\dot{\beta}_{1,0}\dot{\beta}_{1,1} + d_{6,8}\beta_{0,1}\ddot{\beta}_{2,0} + d_{6,9}\beta_{1,0}\ddot{\beta}_{1,1} + d_{6,10}\dot{\beta}_{0,1}\dot{\beta}_{2,0} = 0$$

$$\ddot{\beta}_{1,2} + 2\zeta_{1,2}\omega_{1,2}\,\dot{\beta}_{1,2} + \omega_{1,2}^2\beta_{1,2} + d_{7,1}\beta_{1,0}\beta_{0,1}\,\ddot{\beta}_{0,1} + d_{7,2}\,\ddot{\beta}_{0,1}\,\beta_{1,1}$$

$$+ d_{7,3}\,\dot{\beta}_{1,0}\,\dot{\beta}_{0,1}\beta_{0,1} + d_{7,4}\,\ddot{\beta}_{1,0}\,\dot{\beta}_{0,1}^2 + d_{7,5}\,\ddot{\beta}_{1,0}\beta_{0,2} + d_{7,6}\beta_{0,1}\,\ddot{\beta}_{1,1} \quad , \qquad (6.73)$$

$$+ d_{7,7}\beta_{1,0}\,\ddot{\beta}_{0,2} + d_{7,8}\beta_{1,0}(\dot{\beta}_{0,1})^2 + d_{7,9}\,\dot{\beta}_{0,1}\,\dot{\beta}_{1,1} + d_{7,10}\,\dot{\beta}_{1,0}\,\dot{\beta}_{0,2} = 0$$

$$\ddot{\beta}_{3,0} + 2\zeta_{3,0}\omega_{3,0}\,\dot{\beta}_{3,0} + \omega_{3,0}^2\beta_{3,0} + d_{8,1}\,\ddot{\beta}_{1,0}\beta_{2,0} + d_{8,2}\,\ddot{\beta}_{1,0}\beta_{1,0}^2$$

$$+ d_{8,3}\beta_{1,0}\,\ddot{\beta}_{2,0} + d_{8,4}\,\dot{\beta}_{1,0}^2\,\dot{\beta}_{1,0} + d_{8,5}\,\dot{\beta}_{1,0}\,\dot{\beta}_{2,0} + d_{8,6}\,v_0(t)\cos(\alpha) = 0 \qquad , \qquad (6.74)$$

$$\ddot{\beta}_{0,3} + 2\zeta_{0,3}\omega_{0,3}\,\dot{\beta}_{0,3} + \omega_{0,3}^2\beta_{0,3} + d_{9,1}\,\ddot{\beta}_{0,1}\beta_{0,2} + d_{9,2}\,\dot{\beta}_{0,1}\,\dot{\beta}_{0,2}$$

$$+ d_{9,3}\beta_{0,1}\,\ddot{\beta}_{0,2} + d_{9,4}\,\ddot{\beta}_{0,1}\,\dot{\beta}_{0,1}^2 + d_{9,5}(\dot{\beta}_{0,1})^2\beta_{0,1} + d_{9,6}\,v_0(t)\sin(\alpha) = 0 \qquad , \qquad (6.75)$$

where $\zeta_{i,j}$ is the damping ratio of $(i,j)^{th}$ sloshing mode, $\omega_{i,j}$ is the natural frequency of $(i,j)^{th}$ sloshing mode. The coefficient $d_{i,j}$ is given by:

$$d_{1,1} = \frac{1}{2}E_{1,0} + \frac{\pi^2}{2E_{1,0}} \qquad (6.76)$$

$$d_{1,2} = \frac{\pi^4}{E_{1,0}E_{2,0}} - \frac{\pi^2}{4} \qquad (6.77)$$

$$d_{1,3} = \frac{1}{2}E_{1,0} - \frac{\pi^2}{E_{2,0}} \qquad (6.78)$$

$$d_{1,4} = \frac{\pi^4}{2E_{1,0}E_{1,1}} - \frac{\pi^2}{2} \qquad (6.79)$$

$$d_{1,5} = \frac{1}{2}E_{1,0} \qquad (6.80)$$

$$d_{1,6} = \frac{\pi^4 r^2}{2E_{0,1}E_{1,1}} \qquad (6.81)$$

$$d_{1,7} = \frac{1}{2}E_{1,0} - \frac{\pi^2}{2E_{1,1}} \qquad (6.82)$$

$$d_{1,8} = \frac{\pi^4}{E_{1,0}E_{1,1}} - \pi^2 \qquad (6.83)$$

119

$$d_{1,9} = \frac{\pi^4 r^2}{2E_{0,1}E_{1,1}} + \frac{\pi^2 r^2 E_{1,0}}{2E_{0,1}} - \frac{\pi^4 r^4 E_{1,0}}{2E_{0,1}{}^2 E_{1,1}}$$ (6.84)

$$d_{1,10} = \frac{1}{2}E_{1,0} - \frac{\pi^2}{2E_{1,1}} + \frac{\pi^2 r^2 E_{1,0}}{2E_{0,1}E_{1,1}}$$ (6.85)

$$d_{1,11} = -\frac{4E_{1,0}}{\pi^2 a}$$ (6.86)

$$d_{2,1} = \frac{1}{2}E_{0,1} + \frac{\pi^2 r^2}{2E_{0,1}}$$ (6.87)

$$d_{2,2} = \frac{\pi^4 r^4}{E_{0,1}E_{0,2}} - \frac{\pi^2 r^2}{4}$$ (6.88)

$$d_{2,3} = \frac{1}{2}E_{0,1} - \frac{\pi^2 r^2}{E_{0,2}}$$ (6.89)

$$d_{2,4} = \frac{\pi^4 r^4}{2E_{0,1}E_{1,1}} - \frac{\pi^2 r^2}{2}$$ (6.90)

$$d_{2,5} = \frac{1}{2}E_{0,1}$$ (6.91)

$$d_{2,6} = \frac{\pi^4 r^2}{2E_{1,0}E_{1,1}}$$ (6.92)

$$d_{2,7} = \frac{\pi^4 r^2}{2E_{1,0}E_{1,1}} + \frac{\pi^2 E_{0,1}}{2E_{1,0}} - \frac{\pi^4 E_{0,1}}{2E_{1,0}{}^2 E_{1,1}}$$ (6.93)

$$d_{2,8} = \frac{1}{2}E_{0,1} - \frac{\pi^2 r^2}{2E_{1,1}}$$ (6.94)

$$d_{2,9} = \frac{\pi^4 r^4}{E_{0,1}E_{1,1}} - \pi^2 r^2$$ (6.95)

$$d_{2,10} = \frac{1}{2}E_{0,1} + \frac{\pi^2 E_{0,1}}{2E_{1,0}E_{1,1}} - \frac{\pi^2 r^2}{2E_{1,1}}$$ (6.96)

$$d_{2,11} = -\frac{4E_{0,1}}{r\pi^2 a}$$ (6.97)

$$d_{3,1} = \frac{1}{2}E_{2,0} - \frac{\pi^2}{E_{1,0}} \tag{6.98}$$

$$d_{3,2} = \frac{1}{4}E_{2,0} - \frac{\pi^2}{E_{1,0}} - \frac{\pi^2 E_{2,0}}{4E_{1,0}^{\ 2}} \tag{6.99}$$

$$d_{4,1} = \frac{1}{2}E_{0,2} - \frac{\pi^2 r^2}{E_{0,1}} \tag{6.100}$$

$$d_{4,2} = \frac{1}{4}E_{0,2} - \frac{\pi^2 r^2}{E_{0,1}} - \frac{\pi^2 r^2 E_{0,2}}{4E_{0,1}^{\ 2}} \tag{6.101}$$

$$d_{5,1} = E_{1,1} - \frac{\pi^2}{E_{1,0}} \tag{6.102}$$

$$d_{5,2} = E_{1,1} - \frac{\pi^2 r^2}{E_{0,1}} \tag{6.103}$$

$$d_{5,3} = E_{1,1} - \frac{\pi^2}{E_{1,0}} - \frac{\pi^2 r^2}{E_{0,1}} \tag{6.104}$$

$$d_{6,1} = \frac{\pi^4}{E_{1,0}E_{1,1}} - \pi^2 - \frac{\pi^2 E_{2,1}}{E_{1,0}} + \frac{4\pi^4}{E_{1,0}E_{2,0}} + \frac{\pi^4 r^2}{2E_{1,0}E_{1,1}} \tag{6.105}$$

$$d_{6,2} = \frac{1}{2}E_{2,1} - \frac{\pi^2}{E_{1,0}} \tag{6.106}$$

$$d_{6,3} = \frac{\pi^4 E_{2,1}}{2E_{1,0}^{\ 2}E_{1,1}} - \pi^2 - \frac{3\pi^2 E_{2,1}}{2E_{1,0}} + \frac{4\pi^4}{E_{1,0}E_{2,0}} + \frac{\pi^4 r^2}{2E_{1,0}E_{1,1}} + \frac{\pi^4}{E_{1,0}E_{1,1}} \tag{6.107}$$

$$d_{6,4} = E_{2,1} - \frac{\pi^2 r^2}{E_{0,1}} \tag{6.108}$$

$$\begin{aligned} d_{6,5} = {} & \frac{\pi^4 r^4}{E_{0,1}E_{1,1}} + \frac{2\pi^4 r^2}{E_{0,1}E_{1,1}} + \frac{4\pi^4}{E_{1,0}E_{2,0}} + \frac{\pi^4 r^2}{2E_{1,0}E_{1,1}} + \frac{\pi^4}{E_{1,0}E_{1,1}} \\ & + \frac{\pi^4 r^4 E_{2,1}}{2E_{1,0}E_{0,1}E_{1,1}} - \frac{\pi^2 r^2 E_{2,1}}{2E_{0,1}} - \frac{\pi^2 E_{2,1}}{E_{1,0}} - \frac{\pi^2 r^2}{2} - \pi^2 \end{aligned} \tag{6.109}$$

$$d_{6,6} = \frac{\pi^4 r^2}{E_{0,1}E_{1,1}} - \frac{\pi^2 r^2}{4} - \frac{\pi^2 r^2 E_{2,1}}{4E_{0,1}} + \frac{\pi^4 r^2}{2E_{0,1}E_{1,1}} \qquad (6.110)$$

$$d_{6,7} = \frac{1}{2}E_{2,1} - \frac{\pi^2}{E_{1,0}} - \frac{\pi^2 r^2}{2E_{1,1}} - \frac{\pi^2}{E_{1,1}} - \frac{\pi^2 E_{2,1}}{2E_{1,0}E_{1,1}} \qquad (6.111)$$

$$d_{6,8} = E_{2,1} - \frac{4\pi^2}{E_{2,0}} \qquad (6.112)$$

$$d_{6,9} = \frac{1}{2}E_{2,1} - \frac{\pi^2 r^2}{2E_{1,1}} - \frac{\pi^2}{E_{1,1}} \qquad (6.113)$$

$$d_{6,10} = E_{2,1} - \frac{\pi^2 r^2}{E_{0,1}} - \frac{4\pi^2}{E_{2,0}} \qquad (6.114)$$

$$d_{7,1} = \frac{4\pi^4 r^4}{E_{0,1}E_{0,2}} + \frac{\pi^4 r^2}{2E_{0,1}E_{1,1}} + \frac{\pi^4 r^4}{E_{0,1}E_{1,1}} - \pi^2 r^2 - \frac{\pi^2 r^2 E_{1,2}}{E_{0,1}} \qquad (6.115)$$

$$d_{7,2} = \frac{1}{2}E_{1,2} - \frac{\pi^2 r^2}{E_{0,1}} \qquad (6.116)$$

$$d_{7,3} = \frac{\pi^4}{E_{1,0}E_{1,1}} + \frac{2\pi^4 r^2}{E_{1,0}E_{1,1}} + \frac{4\pi^4 r^4}{E_{0,1}E_{0,2}} + \frac{\pi^4 r^2}{2E_{0,1}E_{1,1}} + \frac{\pi^4 r^4}{E_{0,1}E_{1,1}}$$
$$+ \frac{\pi^4 r^2 E_{1,2}}{2E_{0,1}E_{1,0}E_{1,1}} - \frac{\pi^2 E_{1,2}}{2E_{1,0}} - \frac{\pi^2 r^2 E_{1,2}}{E_{0,1}} - \frac{\pi^2}{2} - \pi^2 r^2 \qquad (6.117)$$

$$d_{7,4} = \frac{\pi^4}{2E_{1,0}E_{1,1}} + \frac{\pi^4 r^2}{E_{1,0}E_{1,1}} - \frac{\pi^2}{4} - \frac{\pi^2 E_{1,2}}{4E_{1,0}} \qquad (6.118)$$

$$d_{7,5} = E_{1,2} - \frac{\pi^2}{E_{1,0}} \qquad (6.119)$$

$$d_{7,6} = \frac{1}{2}E_{1,2} - \frac{\pi^2}{2E_{1,1}} - \frac{\pi^2 r^2}{E_{1,1}} \qquad (6.120)$$

$$d_{7,7} = E_{1,2} - \frac{4\pi^2 r^2}{E_{0,2}} \qquad (6.121)$$

$$d_{7,8} = \frac{4\pi^4 r^4}{E_{0,1}E_{0,2}} + \frac{\pi^4 r^2}{2E_{0,1}E_{1,1}} + \frac{\pi^4 r^4}{E_{0,1}E_{1,1}}$$
$$+ \frac{\pi^4 r^4 E_{1,2}}{2E_{0,1}^{\ 2}E_{1,1}} - \pi^2 r^2 - \frac{3\pi^2 r^2 E_{1,2}}{2E_{0,1}}$$

(6.122)

$$d_{7,9} = \frac{1}{2}E_{1,2} - \frac{\pi^2 r^2}{E_{0,1}} - \frac{\pi^2}{2E_{1,1}} - \frac{\pi^2 r^2}{E_{1,1}} - \frac{\pi^2 r^2 E_{1,2}}{2E_{0,1}E_{1,1}}$$

(6.123)

$$d_{7,10} = E_{1,2} - \frac{\pi^2}{E_{1,0}} - \frac{4\pi^2 r^2}{E_{0,2}}$$

(6.124)

$$d_{8,1} = \frac{1}{2}E_{3,0} - \frac{3\pi^2}{2E_{1,0}}$$

(6.125)

$$d_{8,2} = \frac{3\pi^4}{E_{1,0}E_{2,0}} - \frac{3\pi^2}{8} - \frac{3\pi^2 E_{3,0}}{8E_{1,0}}$$

(6.126)

$$d_{8,3} = \frac{1}{2}E_{3,0} - \frac{3\pi^2}{E_{2,0}}$$

(6.127)

$$d_{8,4} = \frac{6\pi^4}{E_{1,0}E_{2,0}} + \frac{\pi^4 E_{3,0}}{E_{1,0}^{\ 2}E_{2,0}} - \frac{3\pi^2}{4} - \frac{\pi^2 E_{3,0}}{E_{1,0}}$$

(6.128)

$$d_{8,5} = \frac{1}{2}E_{3,0} - \frac{3\pi^2}{2E_{1,0}} - \frac{3\pi^2}{E_{2,0}} - \frac{\pi^2 E_{3,0}}{E_{1,0}E_{2,0}}$$

(6.129)

$$d_{8,6} = -\frac{4E_{3,0}}{9\pi^2 a}$$

(6.130)

$$d_{9,1} = \frac{1}{2}E_{0,3} - \frac{3\pi^2 r^2}{2E_{0,1}}$$

(6.131)

$$d_{9,2} = \frac{1}{2}E_{0,3} - \frac{3\pi^2 r^2}{2E_{0,1}} - \frac{3\pi^2 r^2}{E_{0,2}} - \frac{\pi^2 r^2 E_{0,3}}{E_{0,1}E_{0,2}}$$

(6.132)

$$d_{9,3} = \frac{1}{2}E_{0,3} - \frac{3\pi^2 r^2}{E_{0,2}}$$

(6.133)

$$d_{9,4} = \frac{3\pi^4 r^4}{E_{0,1} E_{0,2}} - \frac{3\pi^2 r^2}{8} - \frac{3\pi^2 r^2 E_{0,3}}{8 E_{0,1}}$$ (6.134)

$$d_{9,5} = \frac{\pi^4 r^4 E_{0,3}}{E_{0,1}{}^2 E_{0,2}} - \frac{3\pi^2 r^2}{4} - \frac{\pi^2 r^2 E_{0,3}}{E_{0,1}} + \frac{6\pi^4 r^4}{E_{0,1} E_{0,2}}$$ (6.135)

$$d_{9,6} = -\frac{4 E_{0,3}}{9 r \pi^2 a}$$ (6.136)

where a is tank length, r is the ratio of the tank length to tank breadth, b is liquid depth from the bottom of the tank to the fluid surface at rest, and $E_{i,j}$ is given by:

$$E_{i,j} = \pi \sqrt{i^2 + (rj)^2} \cdot \tanh(\frac{\pi b}{a} \sqrt{i^2 + (rj)^2}).$$ (6.137)

The $(i,0)$ denotes the transverse mode, $(0,j)$ denotes the longitudinal mode, and the set $(i \neq 0, j \neq 0)$ is the mixed mode. The natural frequency of the sloshing mode is described as:

$$\omega_{i,j}^2 = \frac{g\pi}{a} \sqrt{i^2 + (rj)^2} \cdot \tanh(\frac{\pi b}{a} \sqrt{i^2 + (rj)^2}).$$ (6.138)

The sloshing frequency is dependent on the liquid depth and tank size. However, the tank size provides a larger influence on the sloshing frequency. A trapezoidal-velocity profile (bang-coast-bang acceleration) is applied to move the container. The maximum driving velocity and maximum driving acceleration of the container in the simulation were set to 25 cm/s and 2.5 m/s², respectively. The sloshing measurement position, $(-0.5a, -0.5b)$, was selected at the area of the free surface near the left diagonal corner of the container.

Simulated residual amplitudes induced by various driving distances are shown in Figure 6.31. The driving angle, liquid depth, tank length and breadth were set to 45°, 90 mm, 182 mm and 102 mm, respectively. As the driving distance changes, peaks and troughs appear in all of the transverse, longitudinal, and mixed modes. Peaks and trough arise when the sloshing caused by the acceleration and deceleration are in phase or out of phase. The locations of the peak and trough vary because of difference of the natural frequency of the transverse, longitudinal, and mixed modes. Due to a huge difference between the nonlinear sloshing model and the linear model, container motions cannot excite the mixed mode for the linear sloshing model. However, the mixed-mode response for the nonlinear model will exhibit relatively large amplitude oscillations.

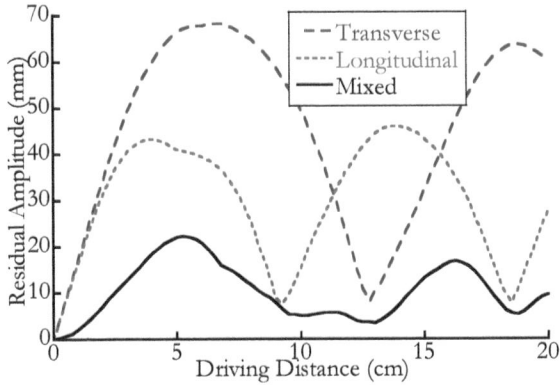

Figure 6.31 Residual amplitudes against driving distance.

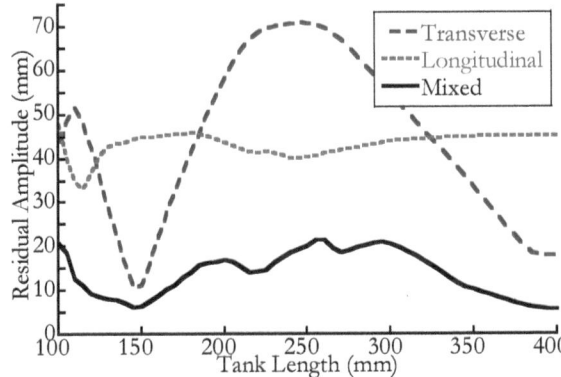

Figure 6.32 Residual amplitudes against container length.

Residual amplitudes induced by various tank lengths are shown in Figure 6.32. The driving distance, steering angle, tank breadth and liquid depth were set to 22.5 cm, 45°, 102 mm, and 90 mm respectively. The residual sloshing in the longitudinal mode varied slightly because of constant tank breadth. Peaks and troughs for transverse and mixed modes occur because of the interference between the sloshing caused by the acceleration and deceleration. Therefore, both the transverse and longitudinal modes are fundamental, while mixed modes still have some effects on the sloshing dynamics.

6.4.2 Simulated Results

Simulated residual amplitudes with the four-pieces smoother induced by various driving distances are shown in Figure 6.33. The steering angle,

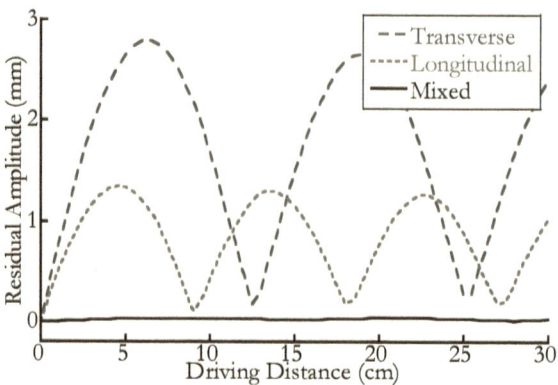

Figure 6.33 Residual amplitudes for various driving distance.

liquid depth, tank length and breadth were set to 45°, 90 mm, 182 mm and 102 mm, respectively. As the driving distance changes, peaks and troughs occur in all of the transverse, longitudinal, and mixed modes. This is because the sloshing caused by the acceleration and deceleration are sometimes in phase and sometimes out of phase. The transverse-mode sloshing was not zero because residual oscillations were designed below the tolerable level for the four-pieces smoother. The dynamics of both the longitudinal mode and mixed mode are similar to that of the transverse mode. With the four-pieces smoother, the residual sloshing of the transverse, longitudinal, and mixed modes was suppressed by an average of 96%, 97%, and 99%, respectively. The four-pieces smoother suppressed the residual sloshing to be a low level for a wide range of driving motions.

Simulated residual amplitudes with the four-pieces smoother induced by various tank lengths ranged from 102 mm to 500 mm are shown in Figure 6.34. The driving distance, steering angle, tank breadth and liquid depth were fixed at 22.5 cm, 45°, 102 mm, and 90 mm respectively. The smoother was designed for the tank length of 300 mm, which is corresponding to the design frequency of 8.7 rad/s. The tank length of 102 mm and 500 mm correspond to the frequency of 17.3 rad/s, and 5.6 rad/s, respectively. For the case of tank length of 300 mm, the four-pieces smoother limited the sloshing of the transverse, longitudinal, and mixed modes to a lower level. For the case of tank length of 102 mm, the four-pieces smoother eliminated the sloshing of the transverse, longitudinal, and mixed modes to <0.1 mm. For the case of 500 mm tank length, the residual amplitude in the transverse mode was large because the four-pieces smoother gets less insensitive at lower frequencies. Therefore, the four-pieces smoother can effectively

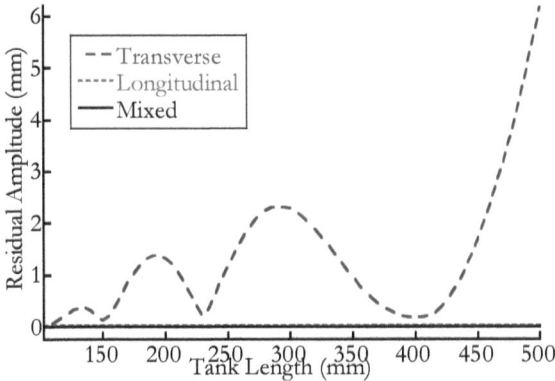

Figure 6.34 Residual amplitudes for various container length.

reduce the total sloshing modes for a large range of working conditions and system parameters.

6.4.3 Experimental Results

Experiments were also conducted on a testing apparatus shown in Figure 6.17. The liquid depth and steering angle were set to 90 mm and 45°, respectively. A camera was also mounted to the gantry for recording the surface elevation at the left diagonal corner of the container. One set of experiments was performed to verify the sloshing dynamics and the effectiveness of the smoother on suppressing nonlinear sloshing. Figure 6.35 shows residual amplitudes of the surface elevation from those tests. As the driving distance increased, experimental residual amplitudes changed. Peaks and troughs occurred because the sloshing caused by the acceleration and deceleration was sometimes in phase and sometimes out of phase. Experimental results match the general trend as simulated curves. Thus, experimental results verified the simulated sloshing dynamics.

The effect of the smoother is also shown in Figure 6.35. Residual amplitudes with the smoother were <3.3 mm for all cases because they were limited to the tolerable level. Experimental data also follow the same general shape as simulated curves. The experiments clearly verified that the four-pieces smoother can effectively eliminate the nonlinear sloshing.

6.5 Chapter Summary

This chapter presented theoretical and experimental analyses of the dynamic effect and control of liquid sloshing for an infinite number of sloshing modes. The two-pieces smoother and three-pieces smoother were

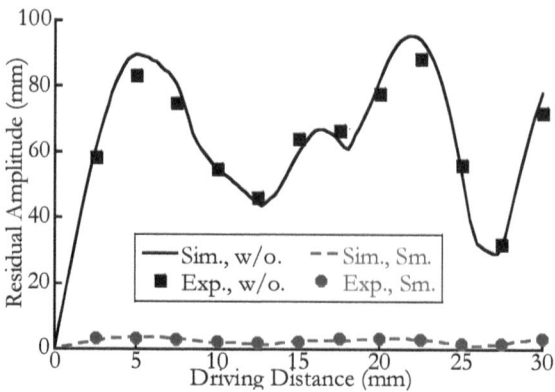

Figure 6.35 Experimental results induced by driving distances.

proposed for suppressing infinite sloshing modes. Both the two- and three-pieces smoothers produced a good control effect that reduced the unwanted transient and residual sloshing. Simulations and experiments were used to verify key dynamic behavior and the effectiveness of the proposed smoothers.

This chapter also presented nonlinear sloshing dynamics in liquid-filled rectangular containers. Operating containers with the addition of nonlinear sloshing is more challenging because of its complicated dynamics. The one-piece smoother and four-pieces smoother were used for sloshing suppression. The smoother provides a robust control effect for various working conditions and system parameters for the nonlinear sloshing in a rectangular container.

Chapter 7. Application in Cam-Follower Systems

7.1 Cam-Follower System

A cam and follower mechanism is widely applied to many types of mechanical machines. The cam and follower mechanisms exhibit excellent properties for operation speed, motion accuracy, structural rigidity and production cost [162-163]. However, the inherent flexibility of the mechanism may induce unwanted vibrations.

Vibrations will reduce position accuracy and cause increased forces, noise, and operating costs [164]. The property of the cam profile directly affects the performance of the cam mechanism. Thus, many works presented numerous cam profiles for vibration reduction including modified s-curves, optimized trigonometric functions and polynomial profiles [165-168]. In many cases, cams need to operate at high operating speeds, which cause the problem of vibration reduction more challenging [169].

Many scientists formulated the problem of vibration reduction in high-speed cam-follower systems over a range of speeds as constrained optimization problems. Jiang et al. [170] designed an environment for designing high-speed cams having a universal Hermite cam displacement. Objective function was residual vibrations, and other cam properties were formulated as design constraints. An optimization program generated computationally the design result. Kwakernaak and Smit [171] presented a linear programming approach and a quadratic method for computing the cam profiles by imposing constraints. Many resulting profiles exhibit small vibrations over a range of cam speeds. Of course, those methods require much computing time and a great memory capacity. Other cam profiles reduced the peak acceleration and jerk with cubic splines [172-173], or adjusted coefficients of polynomials to produce vibration constraints [174-176]. Qiu et al. [177] proposed a universal optimal approach with multiple objectives for either kinematical or dynamic optimization. The resulting profile controlled residual vibrations and limited effectively variations of the velocity and acceleration. By using input shaping techniques, Andresen and Singhose [178] presented modifying high-speed cam profiles. However, derivatives of the cam profile were not demonstrated. In addition, profiles increase the slope of the cam profile at some points, which could cause higher contact forces at the cam surface.

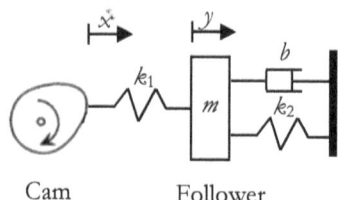

Cam   Follower

Figure 7.1 Cam-follower system.

Rules of thumb and additional constraints have been placed to achieve desired paths for many cam profiles mentioned above. This process increases the complexity of the design. This Chapter is using command smoothing technique to design cam profiles for vibration reduction. Furthermore, the smoothness of the cam profile is developed. Command smoothing is a real-time command modification algorithm for computer-controlled machines. Using knowledge of the natural frequency and damping ratio of the follower, the command smoother is designed to smooth a baseline profile, which will result in vibration reduction and a fast system response [179].

7.2 Dynamics

A typical model of a cam-follower system is illustrated in Figure 7.1 [180]. This model consists of one mass, two springs, and one dashpot. The rotating cam surface profile inputs a desired displacement profile, x, to the follower through a spring, k_1. The mass of the follower denotes as, m. Additional flexibility and damping are modeled by a spring, k_2, and a dashpot, b. Ideally, the follower response, y, tracks the desired displacement of the cam profile, x [181]. Of course, such flexible structures often respond with undesirable vibrations [182].

The equation of the motion for cam-follower systems in Figure 7.1 is:

$$\ddot{y} + 2\zeta\omega\,\dot{y} + \omega^2 y = K\omega^2 x,\tag{7.1}$$

where ω represents the natural frequency of the follower, and ζ is the damping ratio of the follower. They satisfy:

$$\omega = \sqrt{(k_1 + k_2)/m},\tag{7.2}$$

$$\zeta = b/(2m\omega),\tag{7.3}$$

$$K = k_1/(k_1 + k_2).\tag{7.4}$$

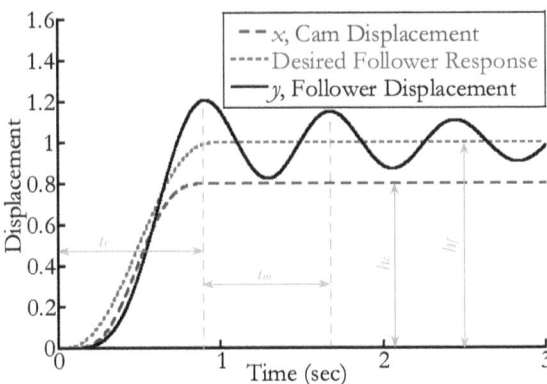

Figure 7.2 Cam and follower displacement.

Figure 7.2 shows the response of the cam and follower. The rise time of the cam is t_r, and the vibration period without damping is t_m, which is $2\pi/\omega$. The maximum desired follower displacement is h_f, and the maximum cam displacement is h_c. The model (7.1) is often written in a normalized coordinate [183]:

$$\ddot{Y} + 2\zeta(2\pi\lambda)\dot{Y} + (2\pi\lambda)^2 Y = (2\pi\lambda)^2 X, \tag{7.5}$$

where the dimensionless displacement of follower represents $Y = y/h_f$, and the dimensionless displacement of the cam is $X = x/h_c$. The speed ratio is defined as $\lambda = t_r/t_m$, and the dimensionless time is $\tau = t/t_r$. Decreasing speed ratio results in an increase in the cam speed.

7.3 Design of Cam Profiles

The cam profile is designed by the smoother, which is described in the Chapter 2. Using the smoother, zero vibration is achieved when the natural frequency, ω, and damping ratio, ζ, of the follower are correct. Baseline profiles filter through the smoother to generate smoothed profiles, which drive the cam-follower system without vibrations. t_b, represents the duration of baseline profiles, and t_s is the duration of the smoother. The sum of the duration of the baseline profile and smoother is the duration of the smoothed profile, t_r. Modifying a basic profile by using the smoother creates a cam profile of rise and dwell stage.

Figure 7.3 shows a process for vibration reduction of the cam and follower system using the command smoothing technique [184]:

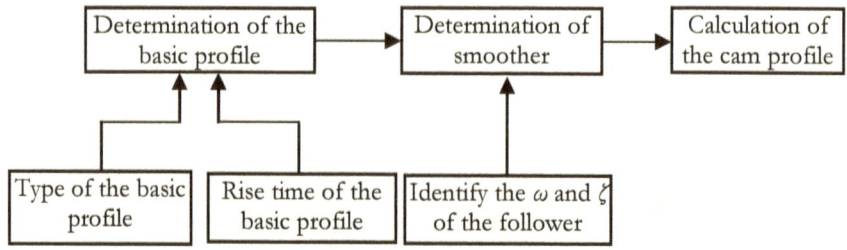

Figure 7.3 Process of cam profile design.

(a) Determination of the basic profile. Basic profile can choose any normal types of cam profiles, which should satisfy the boundary condition of the rise time, $t_b=t_r-t_s$. The duration of the cam profile satisfies: $t_r=\lambda_0 t_m$, where λ_0 is the design operating speed ratio. The speed ratio is often determined by the specific problem considered.

(b) Determination of the smoother. Identifying the natural frequency, ω, and damping ratio, ζ, of the follower produces the smoother.

(c) Calculation of the cam profile. The basic profile is convolved with the smoother to yield the cam profile.

Polynomial profile can be widely used as basic profiles. The general form of a polynomial profile is given by [185]:

$$X(t) = C_0 + C_1 t + C_2 t^2 + \cdots \qquad , \qquad (7.6)$$

where C_n is the polynomial coefficient. The 3-4-5 polynomial profile is a common basic profile, and is named for the order of the term in the polynomial. The 3-4-5 polynomial profile provides continuity for the initial and final conditions of the displacement, velocity, and acceleration for a dwell-rise-dwell motion. When the duration of the 3-4-5 polynomial profile is t_b, the boundary condition satisfy:

$$X(0) = 0, \quad V(0) = 0, \quad A(0) = 0$$
$$X(t_b) = 1, \quad V(t_b) = 0, \quad A(t_b) = 0 \qquad (7.7)$$

Using the boundary condition in equation (7.7), solving equation (7.6) yields the 3-4-5 basic profile:

$$X(t) = 10(t/t_b)^3 - 15(t/t_b)^4 + 6(t/t_b)^5. \qquad (7.8)$$

Convolving the basic profile (7.8) with the one-piece smoother (2.5) obtains the final cam profile. The displacement, velocity, acceleration and jerk properties of the proposed profile are shown in Figure 7.4, where the

132

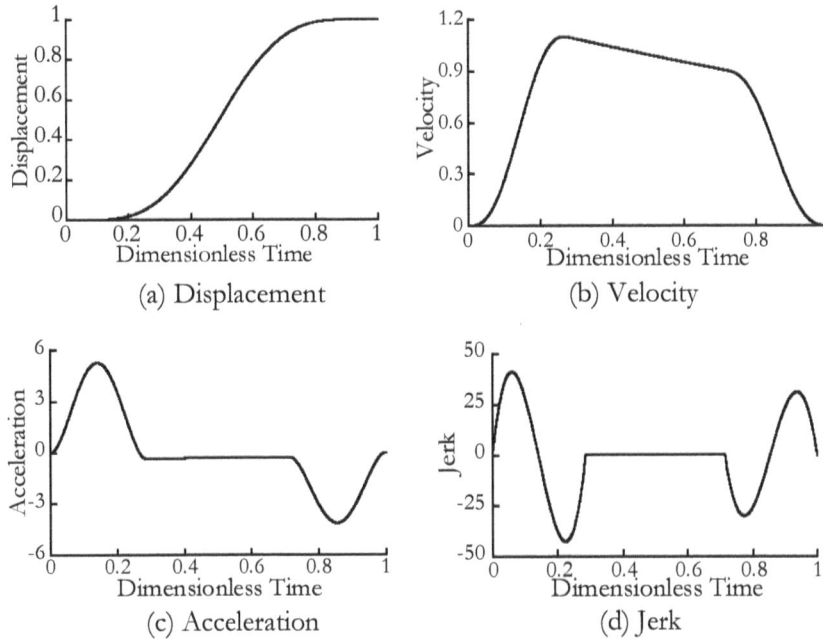

(a) Displacement   (b) Velocity

(c) Acceleration   (d) Jerk

Figure 7.4 Displacement, velocity, acceleration and jerk of the profile.

time is dimensionless. The natural frequency and damping ratio of the follower are set to be 10Hz and 0.05, respectively. The displacement, velocity, acceleration, and jerk of the cam profile are continuous. The continuity improves dynamic performance of a high-speed cam follower system [186]. Therefore, the cam profile provides excellent smoothness.

7.4 Vibration Suppression Properties

Simulated responses of the smoothed and standard 3-4-5 polynomial profile with the same rise time are shown in Figure 7.5. The natural frequency and damping ratio of the follower are also set to 10Hz and 0.05, respectively. Simulated responses show that the new cam profile, which produced by the command smoothing technique, greatly reduces vibrations of the cam follower. Residual vibrations are eliminated by the smoothed profile. However, the standard 3-4-5 polynomial profile with the same rise time will induce larger vibrations when the cam works at high speeds. This is because when the cam operates at high speeds, the rise time of the 3-4-5 polynomial profile will be shorten, which will decrease the low-pass filtering effect.

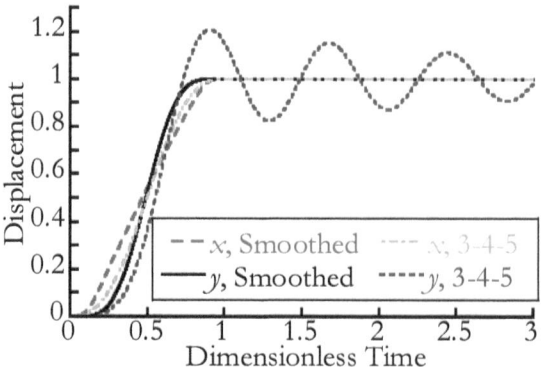

Figure 7.5 Simulated responses to the 3-4-5 polynomial and smoothed profiles.

Simulated results in Figure 7.5 only exhibit the response at a specific operating speed. However, the speed of a cam and follower may be somewhat inconsistent. Therefore, it is important that cams should create low vibrations at the neighborhood of the desired operating speed.

Comparison between the smoothed polynomial profile and standard 3-4-5 polynomial profile over a range of operating speeds is shown in Figure 7.6. Those two type of cam profiles are designed for the speed ratio of 1.3. The horizontal axis is in terms of the real speed ratio, λ. The vibration amplitude is also a normalized value. After the rise stage of the cam profile, the ratio of the maximum vibration amplitude of the follower to the maximum displacement of the follower is defined as the normalized residual amplitude. Zero vibrations with the smoothed cam profile obtained at the design speed ratio of 1.3. When the natural period of the cam follower is held constant, decreasing speed ratio denotes an increase in the cam speed. The comparing results indicate that the smoothed cam profile limits vibrations at a lower level over a large range of speeds. This is because the rise time of the 3-4-5 polynomial profile will be shorten as the cam speed increases, and then the low-pass filtering property will weak.

7.5 Experimental Verification

Experiments were conducted on a rectilinear control plant shown in Figure 7.7 to verify the effectiveness of the proposed solution. A digital signal processor-based motion controller connects a personal computer to a servo amplifier. Cam profiles are issued to the motor through an

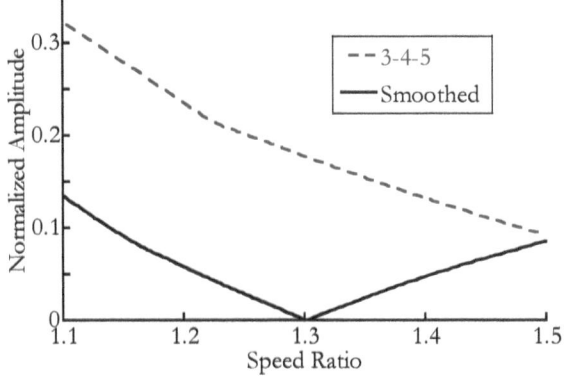

Figure 7.6 Residual amplitude over a range of speed ratio.

Figure 7.7 Rectilinear control plant.

executive program in the personal computer. The mass block 1 connected rigidly with the motor, and the position of the mass block 1 can be measured by the encoder 1. As a result, the displacement of the mass block 1 can trace cam profiles exactly by an advanced feedback controller. The spring 1 and 2, mass block 2 and air dashpot represent the flexibility, follower and damping, respectively. The position of the mass block 2, which represents the follower, can be measured by the encoder 2. The natural frequency and damping ratio of the follower were experimentally measured to be 2.52 Hz and 0.05, respectively.

Experiments were performed to verify the effectiveness of cam profiles for vibration reduction. Cam profiles were issued to the motor when the design speed ratio was set to be 1.3. Simulated and experimental results of normalized residual vibration amplitudes over a range of speed ratios are shown in Figure 7.8. Experimental results agree well with simulated results. Near-zero vibrations for the smoothed cam profile arose at the speed ratio of 1.3 approximately. Around the design speed ratio, the smoothed cam profile exhibited lower vibration than the standard 3-4-5 polynomial profile.

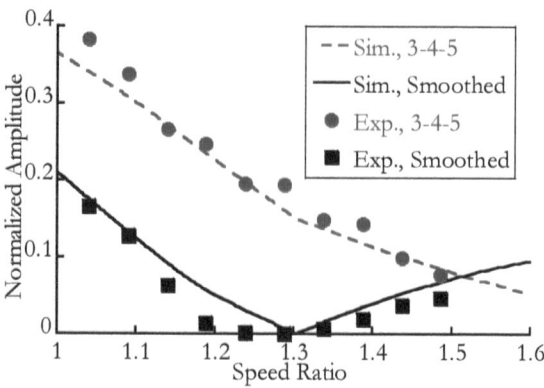

Figure 7.8 Experimental residual amplitudes over a range of speed ratios.

7.6 Chapter Summary

The method of generating high-speed cam profiles for vibration suppression has been presented. The displacement, velocity, acceleration and jerk properties of the smoothed profile demonstrated the profile is smooth, which benefits dynamic performance of high-speed cam-follower systems. A comparison between the proposed profile and 3-4-5 polynomial profile has been conducted. The theoretical analysis showed that the profile obtained by the proposed method can limit vibrations to zero at the design speed. Furthermore, it can limit vibrations below a low level around the design speed. The profile is not complex so that manufacturing cams poses no special problems. Simulations and experimental results verified the effectiveness of the proposed profile for vibration reduction at high and multiple speeds.

Chapter 8. Application in Helicopter Slung Loads

8.1 Introduction

Helicopters carrying large loads are commonly used for transportation and rescue missions. Unfortunately, the slung load attached to the helicopter creates a flexible system, which corrupts the rotorcraft's attitude, and then causes a dangerous operation. The pilot's commanded motions and external disturbances cause unwanted oscillations of the helicopter and load. Therefore, it is essential to study the dynamics and control of this type of aerial cranes for safe and efficient transportation.

Modeling of helicopters carrying slung loads has achieved extensive attention [187-192]. However, much of work is directed at the single-pendulum dynamics, where the load is modeled as a point mass without size. In addition, some previous works are focused on reducing the slung-load swing. Active mechanisms were studied to reduce oscillations [193], but additional mechanisms have increased the complication of the overall system. Meanwhile, the feedback controller uses the measurement of the load oscillations as the control input in a closed loop [194-198]. Regrettably, the feedback controller conflicts with pilot's commands and accurately sensing the slung load is difficult. Input shaping techniques filter the pilot's commands to create low-swing motions [199-201]. The scheme suppresses residual oscillations, but is not effective at reducing transient oscillations.

In many cases, the helicopter hangs a bulky load by cables below the fuselage. However, little attention has been paid to the dynamics and control of helicopters carrying distributed-mass loads. The distributed-mass load dynamics is more complicated than the point-mass dynamics. Because the load swing arises toward the flight direction and the load twisting occurs about the cable, operating helicopters moving large-size loads could be very challenging. Control of distributed-mass load oscillations is very challenging by a factor, which is reducing the load twisting.

8.2 Modeling

A schematic representation of a helicopter transporting a distributed-mass beam is shown in Figure 8.1. The helicopter flies on near-hover conditions along the N_x direction and N_y direction. The mass of the

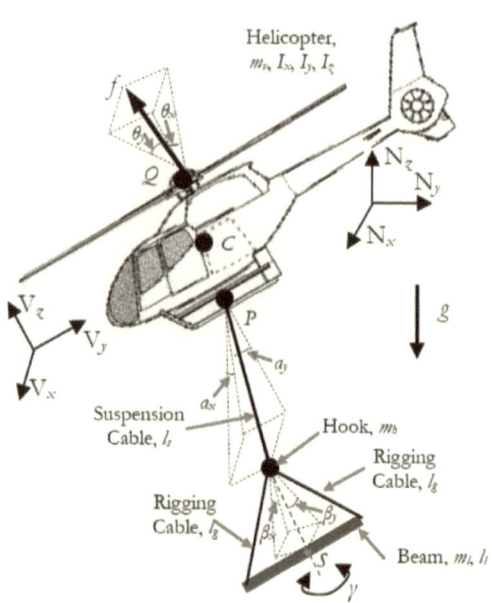

Figure 8.1 Model of a helicopter carrying a distributed-mass beam.

helicopter denotes m_v. The moment of inertia about the N_x, N_y, N_z axis denote I_x, I_y, and I_z, respectively. Motions of the helicopter include the displacement, x, along the N_x direction, the displacement, y, along the N_y direction, the pitch attitude, φ_x, about the N_x axis, and roll attitude, φ_y, about the N_y axis. By rotating the pitch attitude, φ_x, about the N_x axis, and the roll attitude, φ_y, about the N_y axis, the inertial coordinates $N_x N_y N_z$ convert to the moving Cartesian coordinates $V_x V_y V_z$. The coordinates $V_x V_y V_z$ are fixed to the helicopter. The distance between the helicopter center of gravity, C, and the main rotor hub, Q, is a. The angle between the thrust force, f, produced by the main rotor and the $V_y V_z$ plane is defined as the longitudinal rotor angle, θ_x. The angle between the thrust force, f, and the $V_x V_z$ plane is defined as the lateral rotor angle, θ_y.

The distance between the load suspension point, P, and the helicopter center of gravity, C, is b. A suspension cable of length, l_s, suspends below the helicopter and supports a hook of mass, m_h. The cable is assumed to be massless and inelastic. The angle between the suspension cable and the $V_y V_z$ plane is defined as the longitudinal swing angle of the suspension cable, a_x. The angle between the suspension cable and the $V_x V_z$ plane is defined as the lateral swing angle of the suspension cable, a_y. Two rigging cables of length, l_g, connect a uniformly distributed-mass beam of mass, m_l, and length, l_l, to the hook. The centroid of the beam is S. β_x and β_y are angles of the load swing relative to the suspension cable. The angle between the direction of the beam length and the N_y direction is defined as the load twist angle about the rigging cables, γ.

The thrust angles, θ_x, and θ_y, are inputs to the near-hover model. The helicopter displacements, x, and y, the attitudes, φ_x, and φ_y, the swing angles, a_x, a_y, β_x, β_y, and the twist angle, γ, are outputs. The aerodynamic effects on the suspended load are also assumed to be negligible. The kinetic energy of the helicopter is given by:

$$T_V = 0.5[m_v(\dot{x})^2 + m_v(\dot{y})^2 + I_x(\dot{\varphi}_x)^2 + I_y(\dot{\varphi}_y)^2]. \tag{8.1}$$

When the helicopter centroid is set to zero potential energy surface, the potential energy is:

$$V_v = 0. \tag{8.2}$$

The kinetic energy of the hook is given by:

$$
\begin{aligned}
T_h =\ & \left[\begin{array}{l} b_{130}(b_{13}\,\dot{\varphi}_y + b_{14}\,\dot{\varphi}_x + l_s\,\dot{\alpha}_y) + b_1\,\dot{\varphi}_y \\ -b_{132}(b_{17}\,\dot{\alpha}_y + b_{15}\,\dot{\varphi}_x + b_{16}\,\dot{\varphi}_y) - \dot{x} \end{array} \right]^2 \\
& + \frac{m_h}{2} \cdot \left[\begin{array}{l} b_{138}(b_{17}\,\dot{\alpha}_x + b_{15}\,\dot{\varphi}_x + b_{16}\,\dot{\varphi}_y) + b_4 b_{126}\,\dot{\varphi}_x \\ + b_{136}(b_{13}\,\dot{\varphi}_y + b_{14}\,\dot{\varphi}_x + l_s\,\dot{\alpha}_y) - b_{142}b\,\dot{\varphi}_y - \dot{y} \end{array} \right]^2 , \\
& + \frac{m_h}{2} \cdot \left[\begin{array}{l} b_{139}(b_{17}\,\dot{\alpha}_x + b_{15}\,\dot{\varphi}_x + b_{16}\,\dot{\varphi}_y) + b_4 b_{122}\,\dot{\varphi}_x \\ + b_{137}(b_{13}\,\dot{\varphi}_y + b_{14}\,\dot{\varphi}_x + l_s\,\dot{\alpha}_y) + b_{143}b\,\dot{\varphi}_y \end{array} \right]^2
\end{aligned}
\tag{8.3}
$$

where b_i is coefficient. The potential energy of the hook is given by:

$$V_h = (-l_s b_{141} - bb_{145})m_h g, \tag{8.4}$$

where g is the gravitational constant. The kinetic energy of the load is given by:

$$
T_l = \frac{m_h}{2} \cdot \left[\begin{array}{l} b_{149}(b_{114}\,\dot{\alpha}_x - b_{116}\,\dot{\alpha}_y + b_{115}\,\dot{\varphi}_x - b_{117}\,\dot{\varphi}_y) + b_{118}b_{115}\,\dot{\beta}_x - \dot{x} \\ -b_{146}(b_{110}\,\dot{\alpha}_x - b_{112}\,\dot{\alpha}_y + b_{111}\,\dot{\varphi}_x - b_{113}\,\dot{\varphi}_y) + b_{161}\,\dot{\beta}_y l_y + b_1\,\dot{\varphi}_y \\ + b_{130}(b_{13}\,\dot{\varphi}_y + b_{14}\,\dot{\varphi}_x + \dot{\alpha}_y l_s) - b_{132}(b_{17}\,\dot{\alpha}_y + b_{15}\,\dot{\varphi}_x + b_{16}\,\dot{\varphi}_y) \end{array} \right]^2
$$

$$+\frac{m_b}{2}\cdot\left[\begin{array}{l} b_{150}(b_{114}\dot{\alpha}_x - b_{116}\dot{\alpha}_y + b_{115}\dot{\varphi}_x - b_{117}\dot{\varphi}_y) + b_{118}b_{156}\dot{\beta}_x + b_4 b_{126}\dot{\varphi}_x \\ -b_{147}(b_{110}\dot{\alpha}_x - b_{112}\dot{\alpha}_y + b_{111}\dot{\varphi}_x - b_{113}\dot{\varphi}_y) + b_{162}\dot{\beta}_y l_y - b_{142}b\dot{\varphi}_y \\ +b_{138}(b_{17}\dot{\alpha}_x + b_{15}\dot{\varphi}_x + b_{16}\dot{\varphi}_y) + b_{136}(b_{13}\dot{\varphi}_y + b_{14}\dot{\varphi}_x + \dot{\alpha}_y l_s) - \dot{y} \end{array}\right]^2$$

$$+\frac{m_b}{2}\cdot\left[\begin{array}{l} b_{151}(b_{114}\dot{\alpha}_x - b_{116}\dot{\alpha}_y + b_{115}\dot{\varphi}_x - b_{117}\dot{\varphi}_y) + b_{118}b_{157}\dot{\beta}_x + b_{143}b\dot{\varphi}_y \\ -b_{148}(b_{110}\dot{\alpha}_x - b_{112}\dot{\alpha}_y + b_{111}\dot{\varphi}_x - b_{113}\dot{\varphi}_y) + b_{163}\dot{\beta}_y l_y + b_4 b_{122}\dot{\varphi}_x \\ +b_{139}(b_{17}\dot{\alpha}_x + b_{15}\dot{\varphi}_x + b_{16}\dot{\varphi}_y) + b_{137}(b_{13}\dot{\varphi}_y + b_{14}\dot{\varphi}_x + \dot{\alpha}_y l_s) \end{array}\right]^2$$

$$+0.5\left[\begin{array}{l} I_x(b_{20}\dot{\beta}_x + b_{191}\dot{\beta}_y)^2 + I_y(b_{192}\dot{\beta}_y - b_{22}\dot{\beta}_x)^2 \\ +I_z(b_{193}\dot{\beta}_y + \dot{\gamma} - b_{23}\dot{\beta}_x)^2 \end{array}\right].$$

$$(8.5)$$

The potential energy of the load is given by:

$$V_l = (-l_y b_{154} - l_s b_{141} - bb_{145})\cdot m_l g, \qquad (8.6)$$

where

$$l_y = \sqrt{l_l^2 - 0.25 l_s^2}. \qquad (8.7)$$

Therefore, the total kinetic and the potential of the system are:

$$T = T_v + T_b + T_l, \qquad (8.8)$$

$$V = V_v + V_b + V_l. \qquad (8.9)$$

The generalized force in hover operation is expressing as:

$$f = \frac{m_v g + m_l g + m_b g}{\left[\begin{array}{l} -\sin\theta_y \sin\varphi_x \cos\varphi_x - \cos\theta_y \sin\theta_x \sin\varphi_x \\ +\cos\theta_y \cos\theta_x \cos\varphi_y \cos\varphi_x \end{array}\right]}. \qquad (8.10)$$

Using the generalized Lagrange method, equations of the motion can be derived:

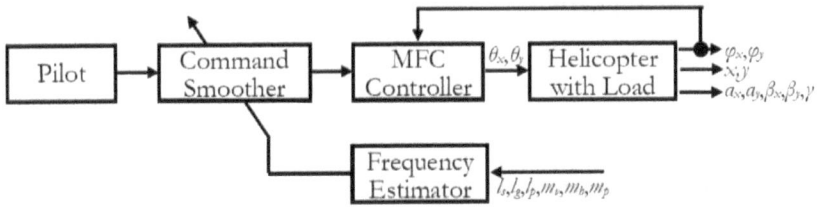

Figure 8.2 Combined control architecture.

$$
\begin{pmatrix}
\ddot{x}_v \\
\ddot{y}_v \\
\ddot{\varphi}_x \\
\ddot{\varphi}_y \\
\ddot{\alpha}_x \\
\ddot{\alpha}_y \\
\ddot{\beta}_x \\
\ddot{\beta}_y \\
\ddot{\gamma}
\end{pmatrix}
= f
\begin{pmatrix}
x_v, y_v, \varphi_x, \varphi_y, \alpha_x, \alpha_y, \beta_x, \beta_y, \gamma, \\
\dot{x}_v, \dot{y}_v, \dot{\varphi}_x, \dot{\varphi}_y, \dot{\alpha}_x, \dot{\alpha}_y, \dot{\beta}_x, \dot{\beta}_y, \dot{\gamma}, \\
\theta_x, \theta_y, \dot{\theta}_x, \dot{\theta}_y, \ddot{\theta}_x, \ddot{\theta}_y
\end{pmatrix}.
\tag{8.11}
$$

8.3 Oscillation Suppression

Oscillations of the helicopter's attitude and the load are controlled by a combined controller including feedback controller and open-loop controller. By following the state of a prescribed model and attenuating the tracking error, a model-following controller (MFC) regulates the helicopter's attitude. By smoothing the pilot's command, a four-pieces smoother reduces both the load swing and twisting. Figure 8.2 shows the combined control architecture. The frequency is estimated by some known system parameters for designing the four-pieces smoother. The pilot's command is convolved with the four-pieces smoother to produce a smoothed command. The smoothed command drives the helicopter toward the desired position with minimum oscillations. The MFC controller measures helicopter's attitudes to adjust the rotor angle in a closed loop.

8.3.1 Attitude Regulation

It is very challenging to design a MFC controller from the complicated dynamics in (8.11). Therefore, a simplified model without the slung load was used to design a MFC controller. Then the design result was refined as needed for helicopter with slung load. To test the performance of the design result, the controller will be applied on the dynamics of helicopter with distributed-mass beam in (8.11). Ignoring the load swing and twisting in (8.11), a simplified model was derived:

$$
\begin{cases}
I_x \ddot{\varphi}_x \cos(\theta_x + \varphi_x) = a \cdot g \cdot m_v \cdot \sin(\theta_x) \\
I_y \ddot{\varphi}_y \cos(\theta_y + \varphi_y) = a \cdot g \cdot m_v \cdot \sin(\theta_y) \\
\ddot{x} = g \cdot \tan(\theta_x + \varphi_x) \\
\ddot{y} = g \cdot \tan(\theta_y + \varphi_y)
\end{cases}
\tag{8.12}
$$

Rotor angles and attitude angles both are assumed to be small. Then equation (8.12) yields a linearized model. From the linearized model, an explicit MFC controller is designed:

$$
\begin{cases}
\ddot{\varphi}_{xm} + 2\zeta_p \omega_p \dot{\varphi}_{xm} + \omega_p^2 \cdot \varphi_{xm} = \omega_p^2 \cdot c_x \\
\ddot{\varphi}_{ym} + 2\zeta_p \omega_p \dot{\varphi}_{ym} + \omega_p^2 \cdot \varphi_{ym} = \omega_p^2 \cdot c_y \\
\theta_x = I_x \ddot{\varphi}_{xm} / a / g / m_v + k_{xd} \cdot (\dot{\varphi}_{xm} - \dot{\varphi}_x) + k_{xp} \cdot (\varphi_{xm} - \varphi_x) \\
\theta_y = I_y \ddot{\varphi}_{ym} / a / g / m_v + k_{yd} \cdot (\dot{\varphi}_{ym} - \dot{\varphi}_y) + k_{yp} \cdot (\varphi_{ym} - \varphi_y)
\end{cases}
\tag{8.13}
$$

where ω_p is the frequency of the prescribed model, ζ_p is the damping ratio of the prescribed model, k_{xp}, k_{xd}, k_{yp}, and k_{yd} are control gains, c_x and c_y are smoothed pilot's commands along the N_x and N_y directions, and φ_{xm} and φ_{ym} are model outputs along the N_x and N_y directions.

In the equation (8.13), first two equations are prescribed model, where the reasonable damping ratio (=0.707) and reasonable settling time (≤2s) are required. Corresponding eigenvalues of the prescribed model are designed to be -2±2i approximately. By the pole placement technique, the frequency, ω_p, of the prescribed model was derived to be 2.83 rad/s.

In the equation (8.13), last two equations are the asymptotic tracking control law. Considering the coupling effect between the helicopter and the slung load, desired poles of the tracking controller are designed to be -18±8.7i [200]. The helicopter mass, m_v, distance, a, moment of inertia, I_x, I_y, in this paper are 6000 kg, 3.5 m, 17450 kgm², 20500 kgm², respectively. Using the pole placement method, control gains, k_{xp}, k_{xd}, k_{yp}, and k_{yd} are calculated to be 33.916, 3.053, 39.845, and 3.586, respectively.

8.3.2 Swing Reduction

Ignoring helicopter's attitudes and displacements in (8.11), another simplified model can be derived to describe the double-pendulum load swing. Especially, the beam length is assumed to coincide with the flight direction, and oscillations are supposed to be small around the equilibrium position. Then linearized frequencies of the double-pendulum swing are given by:

$$\omega_{2,1} = \sqrt{\frac{g \cdot (w \pm v)}{2 l_s}}, \tag{8.14}$$

where

$$w = \frac{1}{e \cdot l_y^2 + (c + e) \cdot R^2} \left(\begin{array}{l} \left(e^2 + e + c + e \cdot c \right) \cdot l_y^2 + (e + c) \cdot l_y \cdot l_s \\ + \left(e^2 + c^2 + 2e \cdot c + c + e \right) \cdot R^2 \end{array} \right), \tag{8.15}$$

$$v = \sqrt{w^2 - \frac{4\left(e^2 + c^2 + 2e \cdot c + e + c \right) \cdot l_y \cdot l_s}{e \cdot l_y^2 + (e + c) \cdot R^2}}, \tag{8.16}$$

$$R = l_p / (2\sqrt{3}), \tag{8.17}$$

e is the ratio of the hook mass to the helicopter mass, and c is the ratio of the load mass to the helicopter mass. The double-pendulum simplified model can be considered as two uncoupled second-order systems with natural frequencies given in equation (8.14). The damping of the load swing can be assumed to be zero.

A four-pieces smoother should be applied to reduce load oscillations. By the estimate of the first-mode frequency of the load swing from equation (8.14), the smoother can be designed. The low-pass filtering effect of the smoother can suppress the high-mode swing. As a result, helicopter's attitudes are regulated by the MFC controller (8.13), and oscillations of the load are attenuated by the four-pieces smoother.

8.4 Numerical Verification

The hook mass, m_h, load mass, m_p, suspension cable length, l_s, rigging cable length, l_g, load length, l_p, distance, b are fixed at 50 kg, 2000 kg, 15 m,

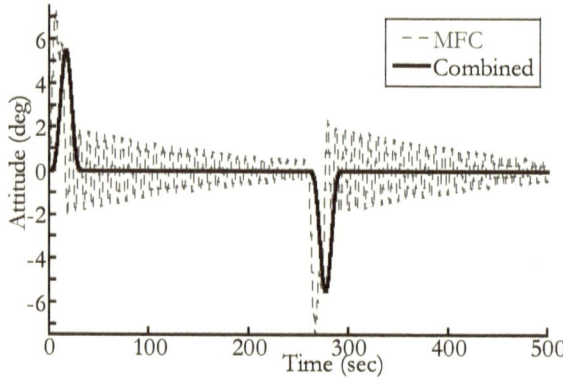

Figure 8.3 Helicopter attitude to a pilot command.

7 m, 10 m, 5 m, respectively. An acceleration was given to move the helicopter flying foreword along the N_y direction, and then the rotorcraft kept the flight speed at 46 km/h. At the end of travel, a deceleration was given to stop the aerial crane at a 3.2 km distance. The normal straight-line flight with suspended loads was simulated in this process. Figures 8.3-8.6 shows the detailed simulated response for the flight distance.

Simulated results of the helicopter's attitude are shown in Figure 8.3. The maximum peak-to-peak amplitudes of the helicopter's attitude with the MFC controller and combined controller were 7.5 degree and 5.5 degree, respectively. The time required for the residual response to settle within 0.5 degree is defined as the settling time of the helicopter's attitude. The settling time with the MFC controller and combined controller were 219.1 seconds and 11.9 seconds, respectively. Therefore, the combined controller obviously reduces more oscillations of the helicopter's attitude. Additionally, the frequency of the helicopter's attitude in Figure 8.3 is less than both the design frequency of the MFC controller and the first-mode frequency of the load swing. This is because the dynamics of the helicopter and slung loads have complex coupling effect, while the two frequencies were estimated from the uncoupled helicopter and slung load.

Simulated results of deflection of load center are shown in Figure 8.4. With the MFC controller, the maximum peak-to-peak amplitude and the settling time of the load deflection were 3.6 m and 173.5 seconds, respectively. The time required for residual oscillations to settle within 1% of the sum of the suspension length and second cable equivalent length is defined as the settling time of the load deflection. With the combined controller, the maximum peak-to-peak amplitude and settling time were 1.8 m and 7.1 seconds, respectively. Therefore, the combined controller also markedly reduces more oscillations of the load swing. In addition, the

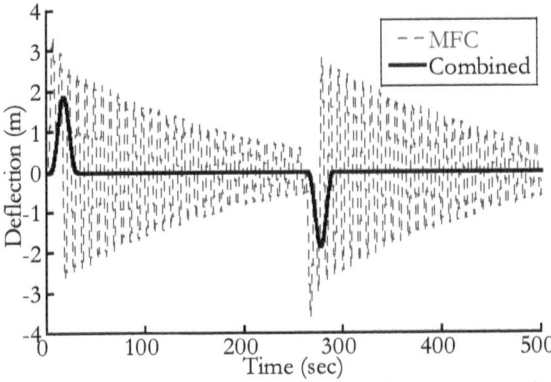

Figure 8.4 Load deflection to a pilot command.

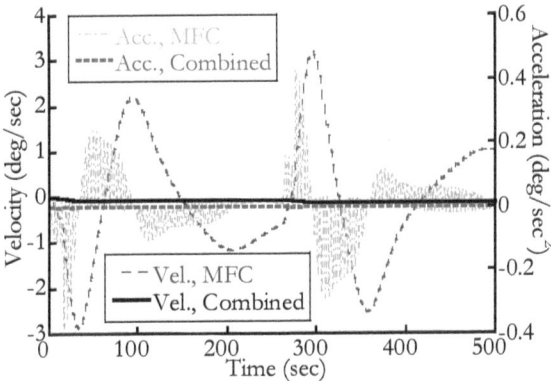

Figure 8.5 Load twisting to a pilot command.

oscillation frequency of the load deflection in Figure 8.4 is similar to the frequency of the helicopter's attitude in Figure 8.3 because of the coupling effect between the helicopter and the load swing.

Simulated results of load twisting is shown in Figure 8.5. With the MFC controller, a trough arose in the angular velocity of the load twisting near 30 seconds, 200 seconds, and 360 seconds. Meanwhile, a peak occurred near 100 seconds, and 300 seconds. This is because oscillations caused by the acceleration and deceleration are sometimes in phase and sometimes out of phase. The velocity would approach a constant after the trough and peak because the frequency of the load twisting depends on the oscillation amplitude of the load swing. The residual velocity with the MFC controller and combined controller were 0.37 degree/second and 0.08 degree/second, respectively. The angular velocity of the load twisting was limited to a small

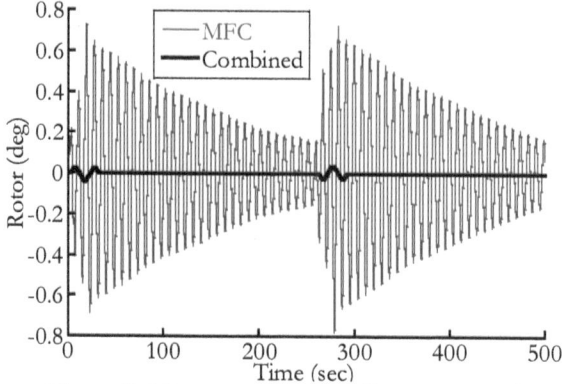
Figure 8.6 Rotor angle to a pilot command.

value by the combined controller for safer operation. Thus, the combined controller also reduces oscillations of the load twisting.

Simulated results of the rotor angle are shown in Figure 8.6. The maximum peak-to-peak rotor angle with the MFC controller and combined controller were 0.78 degree and 0.039 degree, respectively. The rotor frequency in Figure 8.6 was as same as the frequency of the helicopter's attitude in Figure 8.3 and the frequency of the load deflection in Figure 8.4 due to the coupling effect between the helicopter and the load, and the damping effect of attitude oscillations by adjusting the rotor angle.

8.5 Chapter Summary

The dynamic effects and oscillation control of helicopters transporting large-size loads are presented. The chapter derives a dynamic model of a helicopter carrying a distributed-mass beam with double-pendulum dynamics, which includes motions of the helicopter's attitude, load swing, and load twisting. A combined MFC controller and smoother was presented. The smoother reduces the load swing and twisting, while the MFC controller adjusts the helicopter's attitude. Simulations verify the effectiveness of the combined MFC controller and smoother technique.

References

[1] W. SINGHOSE, COMMAND SHAPING FOR FLEXIBLE SYSTEMS: A REVIEW OF THE FIRST 50 YEARS, INTERNATIONAL JOURNAL OF PRECISION ENGINEERING AND MANUFACTURING, VOL. 10, NO. 4, PP.153-168, 2009.

[2] Z. QIU, A DAPTIVE NONLINEAR VIBRATION CONTROL OF A CARTESIAN FLEXIBLE MANIPULATOR DRIVEN BY A BALL SCREW MECHANISM, MECHANICAL SYSTEMS AND SIGNAL PROCESSING, VOL. 30, PP.248–266, 2012.

[3] H. YU, Y. LIN, C. CHU, ROBUST MODAL VIBRATION SUPPRESSION OF A FLEXIBLE ROTOR, MECHANICAL SYSTEMS AND SIGNAL PROCESSING, VOL. 21, PP.334-347, 2007.

[4] D. KIM, W. SINGHOSE, PERFORMANCE STUDIES OF HUMAN OPERATORS DRIVING DOUBLE-PENDULUM BRIDGE CRANES, CONTROL ENGINEERING PRACTICE, VOL. 18, NO. 6, PP.567-576, 2010.

[5] J. LAWRENCE, W. SINGHOSE, COMMAND SHAPING SLEWING MOTIONS FOR TOWER CRANES, JOURNAL OF VIBRATION AND ACOUSTICS, VOL. 132, NO. 1, P.011002, 2010.

[6] D. LEWIS, G. G. PARKER, B. DRIESSEN, ET AL., COMMAND SHAPING CONTROL OF AN OPERATOR-IN-THE-LOOP BOOM CRANE, AMERICAN CONTROL CONFERENCE, PHILADELPHIA, USA, 1998, PP. 2643-2647.

[7] Z. N. MASOUD, M. F. DAQAQ, A GRAPHICAL APPROACH TO INPUT-SHAPING CONTROL DESIGN FOR CONTAINER CRANES WITH HOIST, IEEE TRANSACTIONS ON CONTROL SYSTEMS TECHNOLOGY, VOL. 14, NO. 6, PP. 1070-1077, 2006.

[8] W. SINGHOSE, W. SEERING, N. SINGER, IMPROVING REPEATABILITY OF COORDINATE MEASURING MACHINES WITH SHAPED COMMAND SIGNALS, PRECISION ENGINEERING, VOL. 18, NO. 2-3, PP. 138-146, 1996.

[9] T. D. TUTTLE, W. P. SEERING, VIBRATION REDUCTION IN 0-G USING INPUT SHAPING ON THE MIT MIDDECK ACTIVE CONTROL EXPERIMENT, AMERICAN CONTROL CONFERENCE, SEATTLE, WA, 1995, PP. 919-923.

[10] N. C. SINGER, W. P. SEERING, PRESHAPING COMMAND INPUTS TO REDUCE SYSTEM VIBRATION, JOURNAL OF DYNAMIC SYSTEMS, MEASUREMENT, AND CONTROL, VOL. 112, PP. 76-82, 1990.

[11] T. CHANG, K. GODBOLE, E. HOU, OPTIMAL INPUT SHAPER DESIGN FOR HIGH-SPEED ROBOTIC WORKCELLS, JOURNAL OF VIBRATION AND CONTROL, VOL. 9, NO. 12, PP.1359-1376, 2003.

[12] M. FREESE, E. F. FUKUSHIMA, S. HIROSE, ET AL., ENDPOINT VIBRATION CONTROL OF A MOBILE ENDPOINT VIBRATION CONTROL OF A MOBILE MINE-DETECTING ROBOTIC MANIPULATOR, AMERICAN CONTROL CONFERENCE, NEW YORK, USA, 2007, PP. 7-12.

[13] J. FORTGANG, W. SINGHOSE, J. J. MARQUEZ, ET AL., COMMAND SHAPING FOR MICRO-MILLS AND CNC CONTROLLERS, AMERICAN CONTROL CONFERENCE, PORTLAND, USA, 2005, PP. 4531-4536.

[14] T. CHANG, X. SUN, ANALYSIS AND CONTROL OF MONOLITHIC PIEZOELECTRIC NANO-ACTUATOR, IEEE TRANSACTIONS ON CONTROL SYSTEMS TECHNOLOGY, VOL. 9, NO. 1, PP. 69-75, 2001.

[15] S. S. GÜRLEYÜK, S. CINAL, ROBUST THREE-IMPULSE SEQUENCE INPUT SHAPER DESIGN, JOURNAL OF VIBRATION AND CONTROL, VOL. 13, NO.12, PP. 1807-1818, 2007.

[16] P. H. MECKL, P. B. ARESTIDES, M. C. WOODS, OPTIMIZED S-CURVE MOTION PROFILES FOR MINIMUM RESIDUAL VIBRATION, AMERICAN CONTROL CONFERENCE, PHILADELPHIA, USA, 1998, PP. 2627–2631.

[17] D. SIMON, C. ISIK, OPTIMAL TRIGONOMETRIC ROBOT JOINT TRAJECTORIES, ROBOTICA, VOL. 9, NO. 4, NO. 379-386, 1991.

[18] E. BAYO, B. PADEN, ON TRAJECTORY GENERATION FOR FLEXIBLE ROBOTS, JOURNAL OF ROBOTIC SYSTEMS, VOL. 4, NO. 2, PP. 229-235, 1987.

[19] K. ERKORKMAZ, Y. ALTINTAS, HIGH SPEED CNC SYSTEM DESIGN, PART I: JERK LIMITED TRAJECTORY GENERATION AND QUINTIC SPLINE INTERPOLATION, INTERNATIONAL JOURNAL OF MACHINE TOOLS AND MANUFACTURE, VOL. 41, NO. 9, PP.1323-1345, 2001.

[20] A. I. MAHYUDDIN, A. MIDHA, INFLUENCE OF VARYING CAM PROFILE AND FOLLOWER MOTION EVENT TYPES ON PARAMETRIC VIBRATION AND STABILITY OF FLEXIBLE CAM-FOLLOWER SYSTEMS, JOURNAL OF MECHANICAL DESIGN, VOL. 116, NO. 1, PP. 298-305, 1994.

[21] N. SINGER, W. SINGHOSE, W. SEERING, COMPARISON OF FILTERING METHODS FOR REDUCING RESIDUAL VIBRATION, EUROPEAN JOURNAL OF CONTROL, VOL. 5, PP. 208-218, 1999.

[22] X. XIE, J. HUANG, Z. LIANG, USING CONTINUOUS FUNCTION TO GENERATE SHAPED COMMAND FOR VIBRATION REDUCTION, PROCEEDINGS OF THE INSTITUTION OF MECHANICAL

ENGINEERS, PART I, JOURNAL OF SYSTEMS AND CONTROL ENGINEERING, VOL. 227, NO. 6, PP. 523-528, 2013.

[23] X. XIE, J. HUANG, Z. LIANG, VIBRATION REDUCTION FOR FLEXIBLE SYSTEMS BY COMMAND SMOOTHING, MECHANICAL SYSTEMS AND SIGNAL PROCESSING, VOL. 39, PP. 461-470, 2013.

[24] Q. ZANG, J. HUANG, DYNAMICS AND CONTROL OF THREE-DIMENSIONAL SLOSH IN A MOVING RECTANGULAR LIQUID CONTAINER UNDERGOING PLANAR EXCITATIONS, IEEE TRANSACTIONS ON INDUSTRIAL ELECTRONICS, VOL. 62, NO. 4, PP. 2309 - 2318, 2015.

[25] B. CHEN, J. HUANG, DECREASING INFINITE-MODE VIBRATIONS IN SINGLE-LINK FLEXIBLE MANIPULATORS BY A CONTINUOUS FUNCTION, PROCEEDINGS OF THE INSTITUTION OF MECHANICAL ENGINEERS, PART I, JOURNAL OF SYSTEMS AND CONTROL ENGINEERING, VOL. 23, NO. 6, PP. 436-446, 2017.

[26] BIN CHEN, JIE HUANG, JINCHEN JI, CONTROL OF FLEXIBLE SINGLE-LINK MANIPULATORS HAVING DUFFING OSCILLATOR DYNAMICS, MECHANICAL SYSTEMS AND SIGNAL PROCESSING, VOL.121, PP. 44-57, 2019.

[27] L. RAMLI, Z. MOHAMED, A. ABDULLAHI, ET AL., CONTROL STRATEGIES FOR CRANE SYSTEMS: A COMPREHENSIVE REVIEW, MECHANICAL SYSTEMS AND SIGNAL PROCESSING, VOL. 95, PP.1-23, 2017.

[28] Z. MASOUD, A. NAYFEH, A. AL-MOUSA, DELAYED POSITION-FEEDBACK CONTROLLER FOR THE REDUCTION OF PAYLOAD PENDULATIONS OF ROTARY CRANES, JOURNAL OF VIBRATION AND CONTROL, VOL. 9, NO. 1-2, PP. 257-277, 2003.

[29] Z. MASOUD, M. DAQAQ, N. NAYFEH, PENDULATION REDUCTION ON SMALL SHIP-MOUNTED TELESCOPIC CRANES, JOURNAL OF VIBRATION AND CONTROL, VOL. 10, NO. 8, PP. 1167-1179, 2004.

[30] Z. MASOUD, A. NAYFEH, N. NAYFEH, SWAY REDUCTION ON QUAY-SIDE CONTAINER CRANES USING DELAYED FEEDBACK CONTROLLER: SIMULATIONS AND EXPERIMENTS, JOURNAL OF VIBRATION AND CONTROL, VOL. 11, NO. 8, PP. 1103-1122, 2005.

[31] M. SOLIHIN, W. LEGOWO, A. LEGOWO, FUZZY-TUNED PID ANTI-SWING CONTROL OF AUTOMATIC GANTRY CRANE, JOURNAL OF VIBRATION AND CONTROL, VOL. 16, NO. 1, PP. 127-145, 2010.

[32] G.P. STARR, SWING-FREE TRANSPORT OF SUSPENDED OBJECTS WITH A PATH-CONTROLLED ROBOT MANIPULATOR, JOURNAL OF DYNAMIC SYSTEMS, MEASUREMENT, AND CONTROL, VOL. 107, NO. 1, PP. 97-100, 1985.

[33] K. SORENSEN, W. SINGHOSE, S. DICKERSON, A CONTROLLER ENABLING PRECISE POSITIONING AND SWAY REDUCTION IN BRIDGE AND GANTRY CRANES, CONTROL ENGINEERING PRACTICE, VOL. 15, NO. 7, PP. 825-837, 2007.

[34] J. LAWRENCE, W. SINGHOSE, R. WEISS R, ET AL., AN INTERNET-DRIVEN TOWER CRANE FOR DYNAMICS AND CONTROLS EDUCATION, PROCEEDINGS OF THE 7TH IFAC SYMPOSIUM ADVANCES IN CONTROL EDUCATION, MADRID, SPAIN, 2006, PP. 511-516.

[35] A. ELBADAWY, M. SHEHATA, ANTI-SWAY CONTROL OF MARINE CRANES UNDER THE DISTURBANCE OF A PARALLEL MANIPULATOR, NONLINEAR DYNAMICS, VOL. 82, NO. 1-2, PP. 415-434, 2015.

[36] E. MALEKI, W. SINGHOSE, DYNAMICS AND CONTROL OF A SMALL-SCALE BOOM CRANE, JOURNAL OF COMPUTATIONAL AND NONLINEAR DYNAMICS, VOL. 6, NO. 3, P. 031015, 2011.

[37] G.G. PARKER, K. GROOM, J. HURTADO, ET AL., COMMAND SHAPING BOOM CRANE CONTROL SYSTEM WITH NONLINEAR INPUTS, PROCEEDINGS OF IEEE CONFERENCE OF CONTROL APPLICATIONS, KOHALA COAST, USA, 1999, 1774-1778.

[38] D. BLACKBURN, W. SINGHOSE, J. KITCHEN, ET AL., COMMAND SHAPING FOR NONLINEAR CRANE DYNAMICS, JOURNAL OF VIBRATION AND CONTROL, VOL. 16, NO. 4, PP. 477-501, 2010.

[39] W. SINGHOSE, S. TOWELL, DOUBLE-PENDULUM GANTRY CRANE DYNAMICS AND CONTROL, IEEE CONFERENCE ON CONTROL APPLICATIONS, TRIESTE, ITALY, 1998, PP. 1205-1209.

[40] S. LAHRES, H. ASCHEMANN, O. SAWODNY, ET AL., CRANE AUTOMATION BY DECOUPLING CONTROL OF A DOUBLE PENDULUM USING TWO TRANSLATIONAL ACTUATORS, AMERICAN CONTROL CONFERENCE, CHICAGO, USA, 2000, PP. 1052-1056.

[41] E. ABDEL-RAHMAN, A. NAYFEH, TWO-DIMENSIONAL CONTROL FOR SHIP-MOUNTED CRANES: A FEASIBILITY STUDY, JOURNAL OF VIBRATION AND CONTROL, VOL. 9, NO. 12, PP. 1327-1342, 2003.

[42] W. GUO, D. LIU, J. YI, ET AL., PASSIVITY-BASED-CONTROL FOR DOUBLE-PENDULUM-TYPE OVERHEAD CRANES, IEEE REGION 10 CONFERENCE ON ANALOG AND DIGITAL TECHNIQUES IN ELECTRICAL ENGINEERING, CHIANG MAI, THAILAND, 2004, PP. D546-D549.

148

[43] H. OUYANG, X. DENG, H. XI, ET AL., NOVEL ROBUST CONTROLLER DESIGN FOR LOAD SWAY REDUCTION IN DOUBLE-PENDULUM OVERHEAD CRANES, PROCEEDINGS OF THE INSTITUTION OF MECHANICAL ENGINEERS PART C JOURNAL OF MECHANICAL ENGINEERING SCIENCE, VOL. 233, NO. 12, PP.4359-4371, 2019.

[44] D. LIU, W. GUO, J. YI, GA-BASED COMPOSITE SLIDING MODE FUZZY CONTROL FOR DOUBLE-PENDULUM-TYPE OVERHEAD CRANE, SECOND INTERNATIONAL CONFERENCE ON FUZZY SYSTEMS AND KNOWLEDGE DISCOVERY, CHANGSHA, CHINA, 2005, PP. 792-801.

[45] Z. MASOUD, A. NAYFEH, SWAY REDUCTION ON CONTAINER CRANES USING DELAYED FEEDBACK CONTROLLER, NONLINEAR DYNAMICS, VOL. 34, PP. 347-358, 2003.

[46] Y. AL-SWEITI, D. SÖFFKER, MODELING AND CONTROL OF AN ELASTIC SHIP-MOUNTED CRANE USING VARIABLE GAIN MODEL-BASED CONTROLLER, JOURNAL OF VIBRATION AND CONTROL, VOL. 13, NO. 5, PP.657-685, 2007.

[47] J. VAUGHAN, E. MALEKI, W. SINGHOSE, ADVANTAGES OF USING COMMAND SHAPING OVER FEEDBACK FOR CRANE CONTROL, AMERICAN CONTROL CONFERENCE, BALTIMORE, USA, 2010, PP. 2308-2313.

[48] W. SINGHOSE, D. KIM, M. KENISON, INPUT SHAPING CONTROL OF DOUBLE-PENDULUM BRIDGE CRANE OSCILLATIONS, JOURNAL OF DYNAMIC SYSTEMS, MEASUREMENT, AND CONTROL , VOL. 130, P. 034504, 2008.

[49] H. JAAFAR, Z. MOHAMED, M. SHAMSUDIN, ET AL, MODEL REFERENCE COMMAND SHAPING FOR VIBRATION CONTROL OF MULTIMODE FLEXIBLE SYSTEMS WITH APPLICATION TO A DOUBLE-PENDULUM OVERHEAD CRANE, MECHANICAL SYSTEMS AND SIGNAL PROCESSING, VOL. 115, PP. 677-695, 2019.

[50] R. MANNING, J. CLEMENT, D. KIM, ET AL., DYNAMICS AND CONTROL OF BRIDGE CRANES TRANSPORTING DISTRIBUTED-MASS PAYLOADS, JOURNAL OF DYNAMIC SYSTEMS, MEASUREMENT, AND CONTROL, VOL. 132, P. 014505, 2010.

[51] Z. MASOUD, K. ALHAZZA, E. ABU-NADA, ET AL., A HYBRID COMMAND-SHAPER FOR DOUBLE-PENDULUM OVERHEAD CRANES, JOURNAL OF VIBRATION AND CONTROL, VOL. 20, NO. 1, PP. 24-37, 2014.

[52] J. VAUGHAN, D. KIM, W. SINGHOSE, CONTROL OF TOWER CRANES WITH DOUBLE-PENDULUM PAYLOAD DYNAMICS, IEEE TRANSACTIONS ON CONTROL SYSTEMS TECHNOLOGY, VOL. 18, NO. 6, PP. 1345-1358, 2010.

[53] W. SINGHOSE, D. KIM, MANIPULATION WITH TOWER CRANES EXHIBITING DOUBLE-PENDULUM OSCILLATIONS, IEEE INTERNATIONAL CONFERENCE ON ROBOTICS AND AUTOMATION, ROMA, ITALY, 2007, PP. 4550-4555.

[54] E. MALEKI, W. SINGHOSE, SWING DYNAMICS AND INPUT-SHAPING CONTROL OF HUMAN-OPERATED DOUBLE-PENDULUM BOOM CRANES, JOURNAL OF COMPUTATIONAL AND NONLINEAR DYNAMICS, VOL. 7, P. 031006, 2012.

[55] J. VAUGHAN, J. YOO, N. KNIGHT, ET AL., MULTI-INPUT SHAPING CONTROL FOR MULTI-HOIST CRANES, AMERICAN CONTROL CONFERENCE, WASHINGTON, DC, USA, 2013, PP. 3449-3454.

[56] J. HUANG, E. MALEKI, W. SINGHOSE, DYNAMICS AND SWING CONTROL OF MOBILE BOOM CRANES SUBJECT TO WIND DISTURBANCES, IET CONTROL THEORY AND APPLICATIONS, VOL. 7, NO. 9, PP. 1187-1195, 2013.

[57] J. HUANG, Z. LIANG, Q. ZANG, DYNAMICS AND SWING CONTROL OF DOUBLE-PENDULUM BRIDGE CRANES WITH DISTRIBUTED-MASS BEAMS, MECHANICAL SYSTEMS AND SIGNAL PROCESSING, VOL. 54-55, PP. 357-366, 2015

[58] J. HUANG, X. XIE, Z. LIANG, CONTROL OF BRIDGE CRANES WITH DISTRIBUTED-MASS PAYLOAD DYNAMICS, IEEE/ASME TRANSACTIONS ON MECHATRONICS, VOL. 20, NO. 1, PP. 481-486, 2015.

[59] R. TANG, J. HUANG, CONTROL OF BRIDGE CRANES WITH DISTRIBUTED-MASS PAYLOADS UNDER WINDY CONDITIONS, MECHANICAL SYSTEMS AND SIGNAL PROCESSING, VOL. 72-73, PP. 409-419, 2016.

[60] H. OMAR, A. NAYFEH, GAIN SCHEDULING FEEDBACK CONTROL OF TOWER CRANES WITH FRICTION COMPENSATION, JOURNAL OF VIBRATION AND CONTROL, VOL. 10, NO. 2, PP. 269-289, 2004.

[61] F. RAUSCHER, O. SAWODNY, AN ELASTIC JIB MODEL FOR THE SLEWING CONTROL OF TOWER CRANES, IFAC-PAPERS ONLINE, VOL. 50, NO. 1, PP. 9796-9801, 2017.

[62] T. WU, M. KARKOUB, W. YU, ET AL., ANTI-SWAY TRACKING CONTROL OF TOWER CRANES WITH DELAYED UNCERTAINTY USING A ROBUST ADAPTIVE FUZZY CONTROL, FUZZY SETS AND SYSTEMS, VOL. 290, NO. 1, PP. 118-137, 2016.

[63] R. Samin, Z. Mohamed, Comparative assessment of anti-sway control strategy for tower crane system, AIP Conference Proceedings: Advances in Electrical and Electronic Engineering: From Theory to Applications, vol. 1883, pp. 020035-1-9 2017.

[64] K. Takagi, H. Nishimura, Control of a jib-type crane mounted on a flexible structure, IEEE Transactions on Control Systems Technology, vol. 11, no. 1, pp. 32-42, 2003.

[65] S. Duong, E. Uezato, H. Kinjo, et al., A hybrid evolutionary algorithm for recurrent neural network control of a three-dimensional tower crane, Automation in Construction, vol. 23, pp. 55-63, 2012.

[66] J. Matuško, Š. Ileš, F. Kolonić, et al., Control of 3D tower crane based on tensor product model transformation with neural friction compensation, Asian Journal of Control, vol. 17, no. 2, pp. 443-458, 2015.

[67] H. Sun, Z. Chen, W. Meng, Fuzzy sliding mode anti-swing control for tower crane base on time-delayed filter, Chinese Control and Decision Conference, Taiyuan, China, 2012, pp.2205-2210.

[68] T. Le, V. Dang, D. Ko, et al., Nonlinear controls of a rotating tower crane in conjunction with trolley motion, Proceedings of the Institution of Mechanical Engineers, Part I: Journal of Systems and Control Engineering, vol. 227, no. 5, pp. 451-460, 2013.

[69] I. Carmona, J. Collado, Control of a two wired hammerhead tower crane, Nonlinear Dynamics, vol. 84, no. 4, pp. 2137-2148, 2016.

[70] M. Bock, A. Kugi, Real-time nonlinear model predictive path-following control of a laboratory tower crane, IEEE Transactions on Control Systems Technology, vol. 22, no. 4, pp. 1461-1473, 2014.

[71] A. Le, S. Lee, 3D cooperative control of tower cranes using robust adaptive techniques, Journal of the Franklin Institute, vol. 354, no. 18, pp. 8333-8357, 2017.

[72] N. Sun, Y. Fang, H. Chen, et al., Slew/translation positioning and swing suppression for 4-DOF tower cranes with parametric uncertainties: Design and hardware experimentation, IEEE Transactions on Industrial Electronics, vol. 63, no.10, pp. 6407-6418, 2016.

[73] Š. Ileš, J. Matuško, F. Kolonić, Sequential distributed predictive control of a 3D tower crane, Control Engineering Practice, vol. 79, pp. 22-35, 2018.

[74] J. Wilbanks, C. Adams, M. Leamy, Two-scale command shaping for feedforward control of nonlinear systems, Nonlinear Dynamics, vol. 92, no. 3, pp. 885-903, 2018.

[75] A. Golafshani, J. Aplevich, Computation of time-optimal trajectories for tower cranes, IEEE Conference on Control Applications, Albany, NY, USA, 1995, pp.1134-1139.

[76] A. Tubaileh, Working time optimal planning of construction site served by a single tower crane, Journal of Mechanical Science and Technology, vol. 30, no. 6, pp. 2793-2804, 2016.

[77] W. Book, Recursive lagrangian dynamics of flexible manipulator arms, The International Journal of Robotics Research, vol. 3, no. 3, pp.87-101, 1984.

[78] P.B. Usoro, R. Nadira, S.S. Mahil, A finite element/Lagrange approach to modeling lightweight flexible manipulators, Journal of Dynamic Systems, Measurement, and Control, vol. 108, no. 3, pp. 198-205, 1986.

[79] J.Z. Sasiadek, R. Srinivasan, Dynamic modeling and adaptive control of a single-link flexible manipulator, Journal of Guidance, Control, and Dynamics, vol. 12, no. 6, pp. 838-844, 1989.

[80] T. Yoshikawa, K. Hosoda, Modeling of flexible manipulators using virtual rigid links and passive joints, The International Journal of Robotics Research, vol. 15, no. 3, pp. 290-299, 1996.

[81] D.J. Wagg, S.A. Neild, Nonlinear vibration with control, Springer-Verlag, 2009.

[82] S.M. Kim, Lumped element modeling of a flexible manipulator system, IEEE/ASME Transactions on Mechatronics, vol. 20, no. 2, pp. 967-974, 2015.

[83] A. Walsh, J.R. Forbes, Modeling and control of flexible telescoping manipulators, IEEE Transactions on Robotics, vol. 31, no. 4, pp. 936-947, 2015.

[84] K. Alipour, P. Zarafshan, A. Ebrahimi, Dynamics modeling and attitude control of a flexible space system with active stabilizers, Nonlinear Dynamics, vol. 84, no. 4, pp. 2535-2545, 2016.

[85] S. Choura, A.S. Yigit, Control of a two-link rigid-flexible manipulator with a moving payload mass, Journal of Sound and Vibration, vol. 243, no. 5, pp. 883-897, 2001.

[86] Z. Mohamed, M. Khairudin, A.R. Husain, et al., Linear matrix inequality-based robust proportional derivative control of a two-link flexible manipulator, Journal of Vibration and Control, vol. 22, no. 5, pp. 1244-1256, 2016.

[87] A. Jnifene, Active vibration control of flexible structures using delayed position feedback, Systems & Control Letters, vol. 56, no. 3, pp. 215-222, 2007.

[88] J. Shan, H. Liu, D. Sun, Slewing and vibration control of a single-link flexible manipulator by positive position feedback, Mechatronics, vol. 15, pp. 487-503, 2055.

[89] M. Baroudi, M. Saad, W. Ghie, State-feedback and linear quadratic regulator applied to a single-link flexible manipulator, IEEE International Conference on Robotics and Biomimetics, Guilin, China, 2009, pp. 1381-1386.

[90] J.H. Yang, F.L. Lian, L.C. Fu, Nonlinear adaptive control for flexible-link manipulators, IEEE Transactions on Robotics and Automation, vol. 13, no. 1, pp. 140-148, 1997.

[91] V. Feliu, K.S. Rattan, H.B. Brown, Adaptive control of a single-link flexible manipulator, IEEE Control Systems Magazine, vol. 10, no. 2, pp. 29-33, 1990.

[92] K.S. Yeung, Y.P. Chen, Sliding-mode controller design of a single-link flexible manipulator under gravity, International Journal of Control, vol. 52, no. 1, pp. 101-117, 1990.

[93] S.B. Choi, C.C. Cheong, H.C. Shin, Sliding mode control of vibration in a single-link flexible arm with parameter variations, Journal of Sound and Vibration, vol. 179, no. 5, pp. 737-748, 1995.

[94] L. Tian, C. Collins, Adaptive neuro-fuzzy control of a flexible manipulator, Mechatronics, vol. 15, no. 10, pp. 1305-1320, 2005.

[95] J. Lin, F.L. Lewis, Fuzzy controller for flexible-link robot arm by reduced-order techniques, IEE Proceedings-Control Theory and Applications, vol. 149, no. 3, pp. 177-187, 2002.

[96] M. Isogai, F. Arai, T. Fukuda, Modeling and vibration control with neural network for flexible multi-link structures, IEEE International Conference on Robotics and Automation. Detroit, MI, USA, 1999, pp. 1096-1101.

[97] M. Tinkir, Ü Önen, M. Kalyoncu, Modelling of neurofuzzy control of a flexible link, Proceedings of the Institution of Mechanical Engineers, Part I: Journal of Systems and Control Engineering, vol. 224, no. 5, pp. 529-543, 2010.

[98] A. Mohri, P.K. Sarkar, M. Yamamoto, An efficient motion planning of flexible manipulator along specified path, IEEE International Conference on Robotics and Automation, Leuven, Belgium, 1998, pp. 1104-1109.

[99] A. Abe, Trajectory planning for residual vibration suppression of a two-link rigid-flexible manipulator considering large deformation, Mechanism and Machine Theory, vol. 44, no. 9, pp. 1627-1639, 2009.

[100] H. Kojima, T. Kibe, Optimal trajectory planning of a two-link flexible robot arm based on genetic algorithm for residual vibration reduction, IEEE/RSJ International Conference on Intelligent Robots and Systems, Maui, Hawaii, USA, 2001, 4, pp. 2276-2281.

[101] S. Rhim, W.J. Book, Adaptive time-delay command shaping filter for flexible manipulator control, IEEE/ASME Transactions on Mechatronics, vol. 9, no. 4, pp. 619-626, 2004.

[102] J. Shan, H.T. Liu, D. Sun, Modified input shaping for a rotating single-link flexible manipulator, Journal of Sound and Vibration, vol. 285, no. 1, pp. 187-207, 2005.

[103] S. Kapucu, N. Yildirim, H. Yavuz, et al., Suppression of residual vibration of a translating-swinging load by a flexible manipulator, Mechatronics, vol. 18, no. 3, pp. 121-128, 2008.

[104] Q. Zhang, J.K. Mills, W.L. Cleghorn, et al., Dynamic model and input shaping control of a flexible link parallel manipulator considering the exact boundary conditions, Robotica, vol. 33, no. 6, pp. 1201-1230, 2015.

[105] M. Benosman, G. Le Vey, Control of flexible manipulators: A survey, Robotica, vol. 22, no. 5, pp. 533-545, 2004.

[106] S. Dwivedy, P. Eberhard, Dynamic analysis of flexible manipulators, a literature review, Mechanism and Machine Theory, vol. 41, no. 7, pp.749-777, 2006.

[107] H. RAHIMI, M. NAZEMIZADEH, DYNAMIC ANALYSIS AND INTELLIGENT CONTROL TECHNIQUES FOR FLEXIBLE MANIPULATORS: A REVIEW, ADVANCED ROBOTICS, VOL. 28, NO. 2, PP. 63-76, 2014.

[108] C.T. KIANG, A. SPOWAGE, C.K. YOONG, REVIEW OF CONTROL AND SENSOR SYSTEM OF FLEXIBLE MANIPULATOR, JOURNAL OF INTELLIGENT & ROBOTIC SYSTEMS, VOL. 77, NO. 1, 187-213, 2015.

[109] M. SAYAHKARAJY, Z. MOHAMED, A. FAUDZI, REVIEW OF MODELLING AND CONTROL OF FLEXIBLE-LINK MANIPULATORS, PROCEEDINGS OF THE INSTITUTION OF MECHANICAL ENGINEERS, PART I: JOURNAL OF SYSTEMS AND CONTROL ENGINEERING, VOL. 230, NO. 8, PP. 861-873, 2016.

[110] H. HU, E.H. DOWELL, L.N. VIRGIN, RESONANCES OF A HARMONICALLY FORCED DUFFING OSCILLATOR WITH TIME DELAY STATE FEEDBACK, NONLINEAR DYNAMICS, VOL. 15, NO. 4, PP. 311-327, 1998.

[111] J.C. JI, NONRESONANT HOPF BIFURCATIONS OF A CONTROLLED VAN DER POL-DUFFING OSCILLATOR, JOURNAL OF SOUND AND VIBRATION, VOL. 297, NO. 1, PP. 183-199, 2006.

[112] X. LI, J.C. JI, C.H. HANSEN, ET AL., THE RESPONSE OF A DUFFING-VAN DER POL OSCILLATOR UNDER DELAYED FEEDBACK CONTROL, JOURNAL OF SOUND AND VIBRATION, VOL. 291, NO. 3, PP. 644-655, 2006.

[113] Y. JIN, H. HU, DYNAMICS OF A DUFFING OSCILLATOR WITH TWO TIME DELAYS IN FEEDBACK CONTROL UNDER NARROW-BAND RANDOM EXCITATION, THE TRANSACTIONS OF THE ASME - JOURNAL OF COMPUTATIONAL AND NONLINEAR DYNAMICS, VOL. 3, NO. 2, P. 021205, 2008.

[114] J.C. JI, N. ZHANG, ADDITIVE RESONANCES OF A CONTROLLED VAN DER POL-DUFFING OSCILLATOR, JOURNAL OF SOUND AND VIBRATION, VOL. 315, NO. 1, PP. 22-33, 2008.

[115] C. FENG, W. ZHU, ASYMPTOTIC LYAPUNOV STABILITY WITH PROBABILITY ONE OF DUFFING OSCILLATOR SUBJECT TO TIME-DELAYED FEEDBACK CONTROL AND BOUNDED NOISE EXCITATION, ACTA MECHANICA, VOL. 208, NO. 1-2, PP. 55-62, 2009.

[116] M.S. SIEWE, C. TCHAWOUA, S. RAJASEKAR, PARAMETRIC RESONANCE IN THE RAYLEIGH-DUFFING OSCILLATOR WITH TIME-DELAYED FEEDBACK, COMMUNICATIONS IN NONLINEAR SCIENCE AND NUMERICAL SIMULATION, VOL. 17, NO. 11, PP. 4485-4493, 2012.

[117] Y. WANG, F. LI, DYNAMICAL PROPERTIES OF DUFFING-VAN DER POL OSCILLATOR SUBJECT TO BOTH EXTERNAL AND PARAMETRIC EXCITATIONS WITH TIME DELAYED FEEDBACK CONTROL, JOURNAL OF VIBRATION AND CONTROL, VOL. 21, NO. 2, PP. 371-387, 2015.

[118] S. WEN, Y. SHEN, S.YANG, ET AL., DYNAMICAL RESPONSE OF MATHIEU-DUFFING OSCILLATOR WITH FRACTIONAL-ORDER DELAYED FEEDBACK, CHAOS, SOLITONS & FRACTALS, VOL. 94, PP. 54-62, 2017.

[119] H. YABUNO, BIFURCATION CONTROL OF PARAMETRICALLY EXCITED DUFFING SYSTEM BY A COMBINED LINEAR-PLUS-NONLINEAR FEEDBACK CONTROL, NONLINEAR DYNAMICS, VOL. 12, NO. 3, PP. 263-274, 1997.

[120] M. GHANDCHI-TEHRANI, L. WILMSHURST, S. ELLIOTT, BIFURCATION CONTROL OF A DUFFING OSCILLATOR USING POLE PLACEMENT, JOURNAL OF VIBRATION AND CONTROL, VOL. 21, NO. 14, PP. 2838-2851, 2015.

[121] Y. PENG, J. LI, EXCEEDANCE PROBABILITY CRITERION BASED STOCHASTIC OPTIMAL POLYNOMIAL CONTROL OF DUFFING OSCILLATORS, INTERNATIONAL JOURNAL OF NON-LINEAR MECHANICS, VOL. 46, NO. 2, PP. 457-469, 2011.

[122] F. KHADRA, SUPER-TWISTING CONTROL OF THE DUFFING-HOLMES CHAOTIC SYSTEM, INTERNATIONAL JOURNAL OF MODERN NONLINEAR THEORY AND APPLICATION, VOL. 5, NO. 4, PP.160-170, 2016.

[123] K.S. CHEN, T.S. YANG, K. OU, ET AL., DESIGN OF COMMAND SHAPERS FOR RESIDUAL VIBRATION SUPPRESSION IN DUFFING NONLINEAR SYSTEMS, MECHATRONICS, VOL. 19, NO. 2, PP. 184-198, 2009.

[124] M. GRUNDELIUS, B. BERNHARDSSON, CONTROL OF LIQUID SLOSH IN AN INDUSTRIAL PACKAGING MACHINE, IEEE INTERNATIONAL CONFERENCE ON CONTROL APPLICATIONS, KOHALA COAST, HI, 1999, PP. 1654-1659.

[125] T. ACARMAN, Ü. ÖZGÜNER, ROLLOVER PREVENTION FOR HEAVY TRUCKS USING FREQUENCY SHAPED SLIDING MODE CONTROL, VEHICLE SYSTEM DYNAMICS, VOL. 44, NO. 10, PP. 737-762, 2006.

[126] L. PERTERSON, E. CRAWLEY, R. HANSMAN, NONLINEAR FLUID SLOSH COUPLED TO THE DYNAMICS OF SPACECRAFT, AIAA JOURNAL, VOL. 27, NO. 9, PP. 1230-1240, 1989.

[127] K. MUTO, Y. KASAI, M. NAKAHARA, EXPERIMENTAL TESTS FOR SUPPRESSION EFFECTS OF WATER RESTRAINT PLATES ON SLOSHING OF A WATER POOL, JOURNAL OF PRESSURE VESSEL TECHNOLOGY, VOL. 110, NO. 3, PP. 240-246, 1988.

[128] V. MODI, S. MUNSHI, AN EFFICIENT LIQUID SLOSHING DAMPER FOR VIBRATION CONTROL, JOURNAL OF FLUIDS AND STRUCTURES, VOL. 12, NO. 8, PP. 1055-1071, 1998.

[129] K. BISWAL, S. BHATTACHARYYA, P. SINHA, DYNAMIC CHARACTERISTICS OF LIQUID FILLED RECTANGULAR TANK WITH BAFFLES, JOURNAL OF THE INSTITUTION OF ENGINEERS. INDIA. CIVIL ENGINEERING DIVISION, VOL. 84. PP. 145-148, 2003.

[130] R. VENUGOPAL, D. BERNSTEIN, STATE SPACE MODELING AND ACTIVE CONTROL OF SLOSH, IEEE INTERNATIONAL CONFERENCE ON CONTROL APPLICATIONS, DEARBORN, MI, 1996, PP. 1072-1077.

[131] H. SIRA-RAMIREZ, M. FLIESS, A FLATNESS BASED GENERALIZED PI CONTROL APPROACH TO LIQUID SLOSHING REGULATION IN A MOVING CONTAINER, AMERICAN CONTROL CONFERENCE, ANCHORAGE, AK, 2002, PP. 2909-2914.

[132] B. BANDYOPADHYAY, P. GANDHI, S. KURODE, SLIDING MODE OBSERVER BASED SLIDING MODE CONTROLLER FOR SLOSH-FREE MOTION THROUGH PID SCHEME, IEEE TRANSACTIONS ON INDUSTRIAL ELECTRONICS, VOL. 56, NO. 9, PP. 3432-3442, 2009.

[133] S. KURODE, S. SPURGEON, B. BANDYOPADHYAY, ET AL., SLIDING MODE CONTROL FOR SLOSH-FREE MOTION USING A NONLINEAR SLIDING SURFACE, IEEE/ASME TRANSACTIONS ON MECHATRONICS, VOL. 18, NO. 2, PP. 714-724, 2013.

[134] S. KURODE, B. BANDYOPADHYAY, P.GANDHI, SLIDING MODE CONTROL FOR SLOSH-FREE MOTION OF A CONTAINER USING PARTIAL FEEDBACK LINEARIZATION, INTERNATIONAL WORKSHOP ON VARIABLE STRUCTURE SYSTEMS, ANTALYA TURKEY, 2008, PP. 367-372.

[135] H. RICHTER, MOTION CONTROL OF A CONTAINER WITH SLOSH: CONSTRAINED SLIDING MODE APPROACH, JOURNAL OF DYNAMIC SYSTEMS, MEASUREMENT, AND CONTROL, VOL. 132, NO. 3, P. 031002, 2010.

[136] K. YANO, K. TERASHIMA, ROBUST LIQUID CONTAINER TRANSFER CONTROL FOR COMPLETE SLOSHING SUPPRESSION, IEEE TRANSACTIONS ON CONTROL SYSTEMS TECHNOLOGY, VOL. 9, NO. 3, PP. 483-493, 2001.

[137] K. TERASHIMA, G. SCHMIDT, MOTION CONTROL OF A CART-BASED CONTAINER CONSIDERING SUPPRESSION OF LIQUID OSCILLATIONS, IEEE INTERNATIONAL SYMPOSIUM ON INDUSTRIAL ELECTRONICS, SANTIAGO, CHILE, 1994, PP. 275-280.

[138] M. REYHANOGLU, J. HERVAS, NONLINEAR MODELING AND CONTROL OF SLOSH IN LIQUID CONTAINER TRANSFER VIA A PPR ROBOT, COMMUNICATIONS IN NONLINEAR SCIENCE AND NUMERICAL SIMULATION, VOL. 18, NO. 6, PP. 1481-1490, 2013.

[139] M. REYHANOGLUN, J. HERVAS, NONLINEAR DYNAMICS AND CONTROL OF SPACE VEHICLES WITH MULTIPLE FUEL SLOSH MODES, CONTROL ENGINEERING PRACTICE, VOL. 20, NO.9, PP. 912-918, 2012.

[140] M. GRUNDELIUS, B. BERNHARDSSON, CONSTRAINED ITERATIVE LEARNING CONTROL OF LIQUID SLOSH IN AN INDUSTRIAL PACKAGING MACHINE, PROCEEDINGS OF THE 39TH IEEE CONFERENCE ON DECISION AND CONTROL, SYDNEY, AUSTRALIA, 2000, PP. 4544-4549.

[141] K. YANO, T. TODA, K. TERASHIMA, SLOSHING SUPPRESSION CONTROL OF AUTOMATIC POURING ROBOT BY HYBRID SHAPE APPROACH, PROCEEDINGS OF THE 40TH IEEE CONFERENCE ON DECISION AND CONTROL, ORLANDO, FL, 2001, PP. 1328-1333.

[142] K. YANO, K. TERASHIMA, SLOSHING SUPPRESSION CONTROL OF LIQUID TRANSFER SYSTEMS CONSIDERING A 3-D TRANSFER PATH, IEEE/ASME TRANSACTIONS ON MECHATRONICS, VOL. 20, NO. 1, PP. 8-16, 2005.

[143] Y. NODA, K. YANO, S. HORIHATA, ET AL., SLOSHING SUPPRESSION CONTROL DURING LIQUID CONTAINER TRANSFER INVOLVING DYNAMIC TILTING USING WIGNER DISTRIBUTION ANALYSIS, 43RD IEEE CONFERENCE ON DECISION AND CONTROL, ATLANTIS, BAHAMAS, 2004, VOL. 3, PP. 3045-3052.

[144] P. GANDHI, A. DUGGAL, ACTIVE STABILIZATION OF LATERAL AND ROTARY SLOSH IN CYLINDRICAL TANKS, IEEE INTERNATIONAL CONFERENCE ON INDUSTRIAL TECHNOLOGY, GIPPSLAND, VIC, 2009, PP. 1-6.

[145] J. FEDDEMA, C. DOHRMANN, G. PARKER, ET AL., CONTROL FOR SLOSH-FREE MOTION OF AN OPEN CONTAINER, IEEE CONTROL SYSTEMS, VOL. 17, NO. 1, PP. 29-36, 1997.

[146] S. CHEN, B. HEIN, H. WORN, USING ACCELERATION COMPENSATION TO REDUCE LIQUID SURFACE OSCILLATION DURING A HIGH SPEED TRANSFER, IEEE INTERNATIONAL CONFERENCE ON ROBOTICS AND AUTOMATION, ROMA, ITALY, 2007, PP. 2951-2956.

[147] K. TERASHIMA, M. HAMAGUCHI, K. YANO, MODELING AND INPUT SHAPING CONTROL OF LIQUID VIBRATION FOR AN AUTOMATIC POURING SYSTEM, PROCEEDINGS OF THE 35TH IEEE CONFERENCE ON DECISION AND CONTROL, KOBE, JAPAN, 1996, PP. 4844-4850.

153

[148] B. PRIDGEN, K. BAI, W. SINGHOSE, SHAPING CONTAINER MOTION FOR MULTIMODE AND ROBUST SLOSH SUPPRESSION, JOURNAL OF SPACECRAFT AND ROCKETS, VOL. 50, NO. 2, PP. 440-448, 2013.

[149] K. TERASHIMA, K. YANO, SLOSHING ANALYSIS AND SUPPRESSION CONTROL OF TILTING-TYPE AUTOMATIC POURING MACHINE, CONTROL ENGINEERING PRACTICE, VOL. 9, NO. 6, PP. 607-620, 2001.

[150] M. HAMAGUCHI, Y. YOSHIDA, T. KIHARA, ET AL., PATH DESIGN AND TRACE CONTROL OF A WHEELED MOBILE ROBOT TO DAMP LIQUID SLOSHING IN A CYLINDRICAL CONTAINER, IEEE INTERNATIONAL CONFERENCE ON MECHATRONICS AND AUTOMATION, NIAGARA FALLS, ONT. 2005, VOL. 4, PP. 1959-1964.

[151] N. QI, K. DONG, X. WANG, ET AL., SPACECRAFT PROPELLANT SLOSHING SUPPRESSION USING INPUT SHAPING TECHNIQUE, INTERNATIONAL CONFERENCE ON COMPUTER MODELING AND SIMULATION, MACAU, CHINA, 2009, PP. 162-166.

[152] A. ABOEL-HASSAN, M. ARAFA, A. NASSEF, DESIGN AND OPTIMIZATION OF INPUT SHAPERS FOR LIQUID SLOSH SUPPRESSION, JOURNAL OF SOUND AND VIBRATION, VOL. 320, NO. 1-2, PP. 1-15, 2009.

[153] H. ABRAMSON, THE DYNAMIC BEHAVIOR OF LIQUIDS IN MOVING CONTAINERS, NASA SP-106, 1966.

[154] J. ROBERTS, E. BASURTO, P. CHEN, SLOSH DESIGN HANDBOOK I, NASA TR CR-406, 1966.

[155] F. DODGE, THE NEW DYNAMIC BEHAVIOR OF LIQUIDS IN MOVING CONTAINERS, SOUTHWEST RESEARCH INSTITUTE TECHNICAL REPT. SP-106, SAN ANTONIO, TX, 2000, CHAP. 2.

[156] R. IBRAHIM, V. PILIPCHUK, T. IKEDA, RECENT ADVANCES IN LIQUID SLOSHING DYNAMICS, APPLIED MECHANICS REVIEWS, VOL. 54, NO. 2, P. 133, 2001.

[157] P. GANDHI, K. JOSHI, N. ANANTHKRISHNAN, DESIGN AND DEVELOPMENT OF A NOVEL 2DOF ACTUATION SLOSH RIG, JOURNAL OF DYNAMIC SYSTEMS, MEASUREMENT, AND CONTROL, VOL. 131, NO. 1, P. 011006, 2009.

[158] P. GANDHI, J. MOHAN, K. JOSHI, ET AL., DEVELOPMENT OF 2DOF ACTUATION SLOSH RIG: A NOVEL MECHATRONIC SYSTEM, IEEE INTERNATIONAL CONFERENCE ON INDUSTRIAL TECHNOLOGY, BOMBAY, INDIA, 2006, PP. 1810–1815.

[159] Q. ZANG, J. HUANG, Z. LIANG, SLOSH SUPPRESSION FOR INFINITE MODES IN A MOVING LIQUID CONTAINER, IEEE/ASME TRANSACTIONS ON MECHATRONICS, VOL. 20, NO. 1, PP. 217-225, 2015.

[160] O. FALTINSEN, O. ROGNEBAKKE, A. TIMOKHA, RESONANT THREE-DIMENSIONAL NONLINEAR SLOSHING IN A SQUARE-BASE BASIN, JOURNAL OF FLUID MECHANICS, 487, PP. 1-42, 2003.

[161] J. HUANG, X. ZHAO, CONTROL OF THREE-DIMENSIONAL NONLINEAR SLOSH IN MOVING RECTANGULAR CONTAINERS, THE TRANSACTIONS OF THE ASME - JOURNAL OF DYNAMIC SYSTEMS, MEASUREMENT, AND CONTROL, VOL.140, NO. 8, PP. 081016-8, 2018.

[162] F. FLOCKER, A VERSATILE CAM PROFILE FOR CONTROLLING INTERFACE FORCE IN MULTIPLE-DWELL CAM-FOLLOWER SYSTEMS, JOURNAL OF MECHANICAL DESIGN, VOL. 134, NO. 9, P.094501, 2012.

[163] H. ERDELYI, D. TALABA, A NOVEL METHOD FOR THE DYNAMIC SYNTHESIS OF CAM MECHANISMS WITH AN IMPOSED DRIVING FORCE PROFILE, PROCEEDINGS OF THE INSTITUTION OF MECHANICAL ENGINEERS, PART C: JOURNAL OF MECHANICAL ENGINEERING SCIENCE, VOL. 224, NO. 8, PP.1771-1782, 2010.

[164] J. KUANG, A. LIN, T. HO, DYNAMIC RESPONSES OF A GLOBOIDAL CAM SYSTEM, JOURNAL OF MECHANICAL DESIGN, VOL. 126, NO. 5, PP.909-915, 2004.

[165] L. BIAGIOTTI, C. MELCHIORRI, L. MORIELLO, DAMPED HARMONIC SMOOTHER FOR TRAJECTORY PLANNING AND VIBRATION SUPPRESSION, IEEE TRANSACTIONS ON CONTROL SYSTEMS TECHNOLOGY, DOI: 10.1109/TCST.2018.2882340.

[166] H. LI, M. LE, Z. GONG, ET AL., MOTION PROFILE DESIGN TO REDUCE RESIDUAL VIBRATION OF HIGH-SPEED POSITIONING STAGES, IEEE/ASME TRANSACTIONS ON MECHATRONICS, VOL. 14, NO. 2, PP.264-269, 2009.

[167] M. VÁCLAVÍK, P. JIRÁSKO, RESEARCH AND APPLICATION OF DISPLACEMENT LAWS OF ELECTRONIC CAMS, 12TH IFTOMM WORLD CONGRESS, BESANÇON, FRANCE, JUNE, 2007, PP. 1-6.

[168] W. SINGHOSE, R. ELOUNDOU, J. LAWRENCE, COMMAND GENERATION FOR FLEXIBLE SYSTEMS BY INPUT SHAPING AND COMMAND SMOOTHING, JOURNAL OF GUIDANCE, CONTROL, AND DYNAMICS, VOL. 33, NO. 6, PP.1697-1707, 2010.

154

[169] B. DEMEULENAERE, J. SCHUTTER, ACCURATE REALIZATION OF FOLLOWER MOTIONS IN HIGH-SPEED CAM-FOLLOWER MECHANISMS, PROCEEDINGS OF ISMA, LEUVEN, BELGIUM, SEPTEMBER, 2002, PP. 1107-1106.

[170] J. JIANG, Y. IWAI, H. SU, MINIMIZING AND RESTRICTING VIBRATIONS IN HIGH-SPEED CAM-FOLLOWER SYSTEMS OVER A RANGE OF SPEEDS, JOURNAL OF APPLIED MECHANICS, VOL. 74, NO. 6, PP.1157-1164, 2007.

[171] H. KWAKERNAAK, J. SMIT, MINIMUM VIBRATION CAM PROFILES, JOURNAL OF MECHANICAL ENGINEERING SCIENCE, VOL. 10, NO. 3, PP.219-227, 1968.

[172] K. YOON, S. RAO, CAM MOTION SYNTHESIS USING CUBIC SPLINES, JOURNAL OF MECHANICAL DESIGN, VOL. 115, NO. 3, 441-446, 1993.

[173] E. SANDGREN, R. WEST, SHAPE OPTIMIZATION OF CAM PROFILES USING A B-SPLINE REPRESENTATION, ASME JOURNAL OF MECHANISMS, TRANSMISSIONS AND AUTOMATION IN DESIGN, VOL. 111, PP. 195-201, 2009.

[174] K. SADEK, A. DAADBIN, IMPROVED CAM PROFILES FOR HIGH-SPEED MACHINERY USING POLYNOMIAL CURVE FITTING, PROCEEDINGS OF THE INSTITUTION OF MECHANICAL ENGINEERS, PART E: JOURNAL OF PROCESS MECHANICAL ENGINEERING, VOL. 204, NO. 2, PP.127-132, 1990.

[175] Q. YU, H. LEE, A NEW FAMILY OF PARAMETERIZED POLYNOMIALS FOR CAM MOTION SYNTHESIS, JOURNAL OF MECHANICAL DESIGN, VOL. 117, NO. 4, PP. 653-655, 1995.

[176] M. CHEW, C. CHUANG, MINIMIZING RESIDUAL VIBRATIONS IN HIGH-SPEED CAM-FOLLOWER SYSTEMS OVER A RANGE OF SPEEDS, JOURNAL OF MECHANICAL DESIGN, VOL. 117, PP. 166-172, 1995.

[177] H. QIU, C. LIN, Z. LI, ET AL., A UNIVERSAL OPTIMAL APPROACH TO CAM CURVE DESIGN AND ITS APPLICATIONS, MECHANISM AND MACHINE THEORY, VOL. 40, NO.6, PP.669-692, 2005.

[178] U. ANDRESEN, W. SINGHOSE, A SIMPLE PROCEDURE FOR MODIFYING HIGH-SPEED CAM PROFILES FOR VIBRATION REDUCTION, JOURNAL OF MECHANICAL DESIGN, VOL. 126, PP. 1105-08, 2004.

[179] O. SMITH, FEEDBACK CONTROL SYSTEMS, NEW YORK, MCGRAW-HILL BOOK COMPANY, INC., 1958.

[180] B. FABIEN, R. LONGMAN, F. FREUDENSTEIN, THE DESIGN OF HIGH-SPEED DWELL-RISE-DWELL CAMS USING LINEAR QUADRATIC OPTIMAL CONTROL THEORY, JOURNAL OF MECHANICAL DESIGN, VOL. 116, PP. 867-874, 1994.

[181] B. PRIDGEN, W. SINGHOSE, COMPARISON OF POLYNOMIAL CAM PROFILES AND INPUT SHAPING FOR DRIVEN FLEXIBLE SYSTEMS, JOURNAL OF MECHANICAL DESIGN, VOL. 134, P. 142505, 2012.

[182] Q. YU, H. LEE, INFLUENCE OF CAM MOTIONS ON THE DYNAMIC BEHAVIOR OF RETURN SPRINGS, JOURNAL OF MECHANICAL DESIGN, VOL. 120, NO. 2, PP. 305-310, 1998.

[183] K. KANZAKI, K. ITAO, POLYDYNE CAM MECHANISMS FOR TYPEHEAD POSITIONING, JOURNAL OF ENGINEERING FOR INDUSTRY, VOL. 94, NO. 1, PP.250-254, 1972.

[184] Z. LIANG, J. HUANG, DESIGN OF HIGH-SPEED CAM PROFILES FOR VIBRATION REDUCTION USING COMMAND SMOOTHING TECHNIQUE, PROCEEDINGS OF THE INSTITUTION OF MECHANICAL ENGINEERS, PART C: JOURNAL OF MECHANICAL ENGINEERING SCIENCE, VOL. 228, NO. 18, PP. 3322-3328, 2014.

[185] H. ROTHBART, CAM DESIGN HANDBOOK, NEW YORK: THE MCGRAW-HILL COMPANIES, 2004.

[186] C. UMESH, J. SATISHCHANDRA, SYNTHESIS OF CAM PROFILE USING CLASSICAL SPLINES AND THE EFFECT OF KNOT LOCATIONS ON THE ACCELERATION, JUMP, AND INTERFACE FORCE OF CAM FOLLOWER SYSTEM, JOURNAL OF MECHANICAL ENGINEERING SCIENCE, VOL. 225, PP. 3019-3029, 2011.

[187] W. HALL, A. BRYSON, INCLUSION OF ROTOR DYNAMICS IN CONTROLLER DESIGN FOR HELICOPTERS, JOURNAL OF AIRCRAFT, VOL. 10, NO. 4, PP. 200-206, 1973.

[188] M. BERNARD, J. BENDTSEN, A. LA COUR-HARBO, MODELING OF GENERIC SLUNG LOAD SYSTEM, JOURNAL OF GUIDANCE, CONTROL, AND DYNAMICS, VOL. 32, NO. 2, PP. 573-585, 2009.

[189] M. BERNARD, K. KONDAK, G. HOMMEL, LOAD TRANSPORTATION SYSTEM BASED ON AUTONOMOUS SMALL SIZE HELICOPTERS, AERONAUTICAL JOURNAL, VOL. 114, PP. 191-198, 2010.

[190] K. ENCIU, A. ROSEN, NONLINEAR DYNAMICAL CHARACTERISTICS OF FIN-STABILIZED UNDERSLUNG LOADS, AIAA JOURNAL, VOL. 53, NO. 3, PP.723-738, 2015.

[191] G. GUGLIERI, P. MARGUERETTAZ, DYNAMIC STABILITY OF A HELICOPTER WITH AN EXTERNAL SUSPENDED LOAD, JOURNAL OF THE AMERICAN HELICOPTER SOCIETY, VOL. 59, NO. 4, PP. 1-12, 2014.

[192] Y. CAO, Z. WANG, EQUILIBRIUM CHARACTERISTICS AND STABILITY ANALYSIS OF HELICOPTER SLUNG-LOAD SYSTEM, PROCEEDINGS OF THE INSTITUTION OF MECHANICAL ENGINEERS, PART G: JOURNAL OF AEROSPACE ENGINEERING, VOL. 231, NO. 6, PP.1056-1064, 2016.

[193] T. DUKES, MANEUVERING HEAVY SLING LOADS NEAR HOVER PART I: DAMPING THE PENDULOUS MODE, JOURNAL OF THE AMERICAN HELICOPTER SOCIETY, VOL. 18, NO. 2, PP. 2-11, 1973.

[194] N. GUPTA, A. BRYSON, NEAR-HOVER CONTROL OF A HELICOPTER WITH A HANGING LOAD, JOURNAL OF AIRCRAFT, VOL. 13, PP. 217-222, 1973.

[195] A. HUTTO, FLIGHT-TEST REPORT ON THE HEAVY-LIFT HELICOPTER FLIGHT-CONTROL SYSTEM, JOURNAL OF THE AMERICAN HELICOPTER SOCIETY, VOL. 21, NO. 1, PP. 32-40, 1976.

[196] J. KRISHNAMURTHI, J.F. HORN, HELICOPTER SLUNG LOAD CONTROL USING LAGGED CABLE ANGLE FEEDBACK, JOURNAL OF THE AMERICAN HELICOPTER SOCIETY, VOL. 60, NO. 2, PP. 1-12, 2015.

[197] C. IVLER, CONSTRAINED STATE-SPACE COUPLING NUMERATOR SOLUTION AND HELICOPTER EXTERNAL LOAD CONTROL DESIGN APPLICATION, JOURNAL OF GUIDANCE, CONTROL, AND DYNAMICS, VOL. 38, NO. 10, PP. 2004-2010, 2015.

[198] S. EL-FERIK, A. H. SYED, H. M. OMAR, ET AL., NONLINEAR FORWARD PATH TRACKING CONTROLLER FOR HELICOPTER WITH SLUNG LOAD, AEROSPACE SCIENCE AND TECHNOLOGY, VOL. 69, PP. 602-608, 2017.

[199] M. BISGAARD, A. LA COUR-HARBO, J. BENDTSEN, ADAPTIVE CONTROL SYSTEM FOR AUTONOMOUS HELICOPTER SLUNG LOAD OPERATIONS, CONTROL ENGINEERING PRACTICE, VOL. 18, NO. 7, PP. 800-811, 2010.

[200] C. ADAMS, J. POTTER, W. SINGHOSE, INPUT-SHAPING AND MODEL-FOLLOWING CONTROL OF A HELICOPTER CARRYING A SUSPENDED LOAD, JOURNAL OF GUIDANCE, CONTROL, AND DYNAMICS, VOL. 38, NO. 1, PP. 94-105, 2015.

[201] Y. ZHANG, J. HUANG, J. KATUPITIYA, DYNAMICS AND OSCILLATION CONTROL OF HELICOPTERS CARRYING LARGE-SIZE LOADS, IEEE/ASME INTERNATIONAL CONFERENCE ON ADVANCED INTELLIGENT MECHATRONICS, AUCKLAND, NEW ZEALAND, 2018, PP. 250-255.

www.ingramcontent.com/pod-product-compliance
Lightning Source LLC
Chambersburg PA
CBHW031123180526
45160CB00001B/6